Cold Tolerance in Rice Cultivation

Cold Tolerance in Rice Cultivation

Editor

Pankaj K. Dubey

scitus
academics

Cold Tolerance in Rice Cultivation

Edited by **Pankaj K. Dubey**

Printed in 2017

ISBN: 978-1-68117-022-0

Library of Congress Control Number: 2015931395

SCITUS Academics LLC,
616, Corporate Way, Suite 2, 4766,
Valley Cottage, NY 10989

www.scitusacademics.com

Contents

Preface

As a cereal grain, it is the most widely consumed staple food for a large part of the world's human population, especially in Asia. Thus, investigating the effect of low temperatures is important. The effect of low temperature on rice production are diverse and complex. Since a large portion of maize crops are grown for purposes other than human consumption, rice is the most important grain with regard to human nutrition and caloric intake, providing more than one fifth of the calories consumed worldwide by humans. The plant has many different variations, but is generally a short living plant, with an average life span of 3-7 months, depending on the climate and the variety. It is not a water plant but substantial amounts of water are required for the planting. Cultivated species of rice are considered to be semi-aquatic annuals. The height of the plant can range from 0.4m to over 5m in some floating rices. Rice is cultivated in many different ways around the world, the different methods used can differ greatly even in the same locality. Methods of growing differ greatly in different localities, but in most Asian countries the traditional hand methods of cultivating and harvesting rice are still practiced. The fields are prepared by plowing, fertilizing, and smoothing. The seedlings are started in seedling beds and, after 30 to 50 days, are transplanted by hand to the fields, which have been flooded by rain or river water. During the growing season, irrigation is maintained by dike-controlled canals or by hand watering. The fields are allowed to drain before cutting. In the United States and in many parts of Europe, rice cultivation has undergone the same mechanization at all stages of cultivation and harvesting as have other grain crops. This book provides a clear understanding of cold stress in rice in the hopes that it will provide insight to the subject for further research so that rice plants may be grown efficiently in cold regions with high productivity.

Editor

Chapter 1

Marker-Assisted Breeding of Indonesia Local Rice Variety Siputeh for Semi-Dwarf Phonetype, Good Grain Quality and Disease Resistance to Bacterial Blight

Yanchang Luo[1,3], Sabaruddin Zakaria[2], Bakhtiar Basyah[2], Tingchen Ma[3], Zefu Li[3], Jianbo Yang[3] and Zhongchao Yin[1,4]

[1]Temasek Life Sciences Laboratory, 1 Research Link, National University of Singapore, Singapore 117604, Republic of Singapore

[2]Department of Agrotechnology, Agriculture Faculty, Syiah Kuala University, Darussalam-Banda 23111, Aceh, Indonesia

[3]Rice Research institute, Anhui Academy of Agricultrural Sciences, Hefei 230031, China

[4]Department of Biological Sciences, 14 Science Drive, National University of Singapore, Singapore 117543, Republic of Singapore

ABSTRACT

Background

Rice is one of the most important staple food crops in Asia. Since the first green revolution beginning in 1960s, high-yield semidwarf modern rice varieties have been widely planted; however, traditional rice varieties with tall plant type are still grown in many countries due to their good grain quality and adaptation to local climate and environment. Siputeh, a local rice variety mainly planted in Java and Sumatra islands of Indonesia, produces long grain rice with good cooking and eating quality. However, the variety has low yield with tall plant type and long growth duration and is highly susceptible to biotic and abiotic stress.

Results

Siputeh as the recurrent female was crossed with the donor line WH421, an elite paternal line of hybrid rice containing the *sd1*, Wx^b, *Xa4* and *Xa21* genes, followed by backcrossing and self-pollination. TS4, a BC3F4 line derived from the breeding program, was obtained through marker-assisted selection for the *sd1*, Wx^b, *Xa4* and *Xa21* loci. TS4 has semi-dwarf phenotype and short growth duration. TS4 conferred disease resistance to multiple *Xanthomonas oryzae* pv. *oryzae*(*Xoo*) strains collected from different countries around the world. TS4 achieved higher grain yield than Siputeh in two field trials conducted in Banda Aceh, Indonesia and Lingshui, China, respectively. Finally, TS4 has better grain quality than Siputeh in terms of degree of chalkiness and amylose content.

Conclusion

An improved rice line, designed as TS4, has been developed to contain semi-dwarf gene *sd1*, low amylase content gene Wx^b and bacterial light resistance genes *Xa4* and *Xa21* through marker-assisted selection. TS4 has semi-dwarf phenotype with reduced growth duration, produces high yield with good grain quality and provides broad-spectrum resistance to *Xoo* strains. The development of TS4 enriches the diversity of local rice varieties with high yield potential and good grain quality.

BACKGROUND

Rice (*Oryza sativa* L.) is a staple food crop in most of the Asia countries. The introduction of semi-dwarf rice led to record yield increases throughout Asia in the 1960s. The recessive semi-dwarfing gene, *sd-1*, is one of the most important genes deployed in modern rice breeding. Rice varieties with the *sd1* gene have a shortened culm with improved lodging resistance and a greater harvest index, allowing for the increased use of nitrogen fertilizers (Jennings [1964]). The *sd1* gene has been cloned from rice and its dominant wild-type allele encodes a gibberellin (GA)-20 oxidase-2 (GA20ox-2) (Ashikari et al. [2002]; Sasaki et al. [2002]; Spielmeyer et al. [2002]). The deletion or substitution mutation of the *Os20ox2* gene results in loss of function of the enzyme, which ultimately leads to a deficiency in plant growth hormones gibberellins that control plant height (Ashikari et al. [2002]; Sasaki et al. [2002]; Spielmeyer et al. [2002]). The different alleles of the*sd1* gene have remained the predominant semi-dwarf genes in modern rice cultivars (Asano et al.[2007]).

There are two different types of starch, amylose and amylopectin, in rice endosperm and the amylose content affects rice eating and cooking quality (Cruz and Khush [2000]). With content ranging from 5% to 15%, the low-amylose rice is characterized by fluffy texture, glossy appearance of the cooked rice, soft texture of cooled rice, and excellent expansibility for food processing (Zhu et al. [2003]). Genetic studies have shown that the rice *Wx* gene determines the amylose content in the endosperm. *Wx* encodes granule-bound starch synthase I (GBSSI), a key enzyme for amylose biosynthesis in the endosperm, and *wx* mutant endosperm contains almost exclusively amylopectin (Hori et al. [2007]; Sano [1984]; Wang et al. [1995]). In addition to controlling amylose content, *Wx* also affects gel consistency and gelatinization temperature of rice starch (Su et al. [2011]). Wx^a and Wx^b are two predominantly distributed functional *Wx* alleles in rice (Sano [1984]; Wang et al. [1995]). Wx^a is widely distributed in *indica* rice with high amylose content while Wx^b is mainly found in *japonica* rice with low or intermediate amylose content (Sano[1984]; Wang et al. [1995]). A single nucleotide substitution (G-to-T) at the splice donor site of the first intron in *Wx*^{b} potentially alters the splicing site and decrease the splicing efficiency of the first intron of *Wx*, resulting in the low level of both mature transcript of *Wx*^{b} and amylose content (Cai et al. [1998]; Larkin and Park [1999]; Tian et al. [2009]; Wang et al. [1995]). Based on the polymorphism at the splicing site of the first intron, a co-dominant CAPS (cleaved amplified polymorphic sequence) molecular marker PCR-*Acc*I was developed for selection of *Wx* alleles in rice breeding (Cai et al. [2002]).

Bacterial blight of rice, caused by *Xanthomonas oryzae* pv. *oryzae* (*Xoo*), is one of the most important bacterial diseases prevalent throughout the world (Gnanamanickam et al. [1999]). Rice yield losses caused by bacterial blight can reduce yield by 20-30% and by up to 50% in some areas of Asia (Mew et al. [1993]). The utilization of host disease resistance (*R*) genes is the most economic and efficient method for controlling the disease. *Xa4* is a dominant *R* gene that provides durable resistance to bacterial blight. Cultivars with *Xa4* conferred resistance to almost all Chinese patho-types of *Xoo* except for patho-type C5 (Zhang [2009]). *Xa21* is another dominant bacterial blight gene, which was originally discovered in wild rice specie *Oryza longistaminata* (Ikeda et al.[1990]). IRBB21, an *Xa21* line in IR24 genetic background, showed resistance to all the known races of *Xoo* collected from India and the Philippines (Ikeda et al. [1990]). Both *Xa4* and *Xa21*genes are widely exploited *R* genes in Asian rice breeding programs for bacterial blight resistance (Datta et al. [2002]; Huang et al. [1997]; Luo et al. [2012]; Luo and Yin [2013]; Singh et al.[2001]; Suh et al. [2013]; Zhang et al. [2006]).

Although modern rice varieties have been planted worldwide, elite local rice varieties are still popular in many rice growing countries due to their good grain quality and adaptation to the local climate and environment. Cultivar Siputeh is a local rice variety still planted in Java and Sumatra islands of Indonesia. Siputeh produces long grain rice with good cooking and eating quality. However, the variety has low yield with tall plant type and long growth duration and is highly susceptible to biotic and abiotic stress (Suhartini [1991]). Here we report the genetic improvement of Siputeh for semi-dwarf phenotype, good grain quality and disease resistance to bacterial blight through marker-assisted breeding.

RESULTS

Breeding of TS4

Marker-assisted breeding was conducted between cultivar Siputeh as the recurrent female and cultivar WH421 as the donor line for the *sd1*, *Wx*^{b}, *Xa4* and *Xa21* genes (Figure 1). The genotypes at the *sd1, Wx, Xa4* and *Xa21* loci in each plant at different generations were determined with molecular markers as described in Table 1. In summary, 10 F1 plants, 3 BC1F1 plants, 2 BC2F1 plants and 3 BC3F1 were identified from F1, BC1F1, BC2F1 and BC3F1 generations, respectively (Figure 1). All these plants carried heterozygous alleles at the *sd1*, *Wx*^{b}, *Xa4* and *Xa21* loci (Figure 1). One of the 3 BC3F1 plants (BC3F1 #6) was selected for self-pollination to generate a BC3F2 population, which consisted of 74 BC3F2 individuals. Three BC3F2 plants were identified from the BC3F2 population and they all carried homozygous alleles at the *sd1* and *Xa21* loci and heterozygous alleles at the *Wx*^{b} and *Xa4* loci (Figure 1). Plant BC3F2 #38 was selected to produce a BC3F3 population, which consisted of 146 BC3F3 individuals. Nine plants (BC3F3 #1, 3, 4, 6–11) were identified from the BC3F3 population and they all carried homozygous alleles at the *sd1*, *Wx*^{b}, *Xa4* and *Xa21* loci (Figures 1 and 2). Plant BC3F3 #1 was selected to produce a BC3F4 population. The 48 individual plants in the BC3F4 population showed uniform morphological phenotype and similar growth duration. Plant BC3F4 #30 was designated as TS4 and selected for disease evaluation for resistance to bacterial blight and field trials (Figure 1).

Siputeh(SdlSdl, Wx^aWx^a, xa4xa4, xa21xa21) × WH421(sdlsdl, Wx^bWx^b, Xa4Xa4, Xa21Xa21)

↓

Siputeh(SdlSdl, Wx^aWx^a, xa4xa4, Xa21xa21) × Fl(Sdlsdl, Wx^aWx^b, Xa4Xa4, Xa4xa4, Xa21xa21)

↓

Siputeh(SdlSdl, Wx^aWx^a, xa4xa4, xa21xa21) × BClFl(SdlSdl, Wx^aWx^b, Xa4xa4, Xa21xa21), 3 / 25

↓

Siputeh(SdlSdl, Wx^aWx^a, xa4xa4, xa21xa21) × BC2Fl(Sdlsdl, Wx^aWx^b, Xa4xa21), 2 / 18

↓

BC3Fl(Sdlsdl, Wx^aWx^b, Xa4xa4, Xa21Xa21), 3 / 25

↓

BC3Fl(Sdlsdl, Wx^aWx^b, Xa4xa4, Xa21xa21), 3 / 35

↓ Selfing

BC3F3(sdlsdl, Wx^bWx^b, Xa4Xa4, Xa21Xa21), 9 / 146

↓ Selfing

BC3F4(sdlsdl, Wx^bWx^b, Xa4Xa4, Xa21Xa21Xa21), 1 / 48

↓

BC3F4 #30(sdlsdl, Wx^bWx^b, Xa4Xa4, Xa21Xa21)
(TS4)

Figure 1: Marker-assisted breeding of TS4. The number of positive plants over the number of total plants screened for presence of molecular makers are indicated in the generation undergone marker-assisted selection. *Sd1* and *sd1*, the dominant and recessive alleles of the *Sd1*gene; Wx^a and Wx^b, the alleles *a* and *b* of the rice *Wx* gene; *Xa4* and *xa4*, dominant and recessive alleles of the *Xa4* gene, respectively; *Xa21* and *xa21*, dominant and recessive alleles of the *Xa21* gene, respectively.

Table 1: Molecular markers used in this study

Marker	Gene-of-interest	DNA sequence	Type of marker	Reference
sd1STS	*sd1*	F: 5'ACAAATACCCCAC-CCTCCTGCC3'	STS, co-dominant	(Luo and Yin [2013])
		R: 5'TAGTAGCCTCG-CTCCACGCCC3'		
21	*Xa21*	F: 5'ATAGCAACTG-ATTGCTTGG3'	STS, co-dominant	(Chen et al. [2000])
		R:5' GATCGGTATAA-CAGCAAAAC3'		
RM224[a]	*Xa4*	F:5' ATCGATCGAT-CTTCACGAGG3'	SSR, co-dominant	(Sun et al. [2003])

		R:5′ TGCTATAAAAG-GCATTCGGG3′		
PCR-*Acc*I[b]	*Wx*	F: 5′GCTTCACTTC-TCTGCTTGTG3′	CAPS, co-dominant	(Cai et al. [2002])
		R: 5′ATGATTTAACG-AGAGTTGAA3′		

F, forward primer; R, reverse primer; STS, sequence-tagged site; SSR, simple sequence repeat; CAPS, Cleaved amplified polymorphic sequence.

[a]The genetic distance between RM224 and the *Xa4* locus is 1.1 cM.

[b]CAPS marker at the *Wx* locus by digesting PCR product with *Acc*I to show polymorphism.

Luo *et al.*

Luo *et al. Rice* 2014 **7**:33 doi:10.1186/s12284-014-0033-2

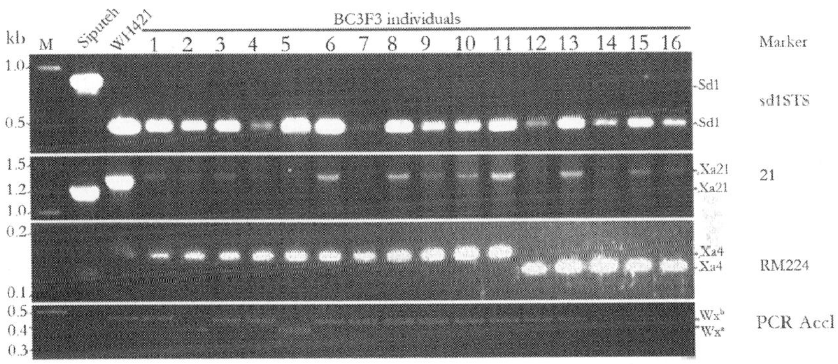

Figure 2: Detection of molecular markers at the*sd1*,*Xa4*, *Xa21*and*Wx*bloci in the B3F3 individuals. The BC3F3 individuals were derived from the cross between Siputeh (recurrent female) and WH421 followed by backcrossing for 3 generations. The amplified PCR products were separated on a 1.5% w/v agarose gel for Marker sd1STS, a 2.0% agarose gel for Marker 21, and a 3.5% agarose gel for Marker PCR-*Acc*I and RM224.BC3F3 plants #1, 3, 4, 6–11 carry homozygous alleles at the *sd1*, *Xa4*, *Xa21* and Wx^b loci.

Luo *et al. Rice* **2014 7**:33 doi:10.1186/s12284-014-0033-2

Disease Evaluation of TS4 for Bacterial Blight Resistance

Twenty-eight *Xoo* strains were used to inoculate WH421, Siputeh and TS4. Siputeh was susceptible to all *Xoo* strains tested while WH421 conferred resistance or moderate resistance to 22 strains and was susceptible or moderate susceptible to 1947, GD1358, A3857, JW89011, K202 and PXO99 (Table 2). TS4 provided resistance or moderate resistance to 25 of the 28 *Xoo* strains tested and was only susceptible to A3857, JW89011 and K202 (Table 2). Compared to WH421, TS4 also provided enhanced resistance with shorter disease lesions to 1947, GD1358, A3857, JW89011, K202 and PXO99 (Table 2). It should be noted that neither WH421 nor TS4 could provide complete resistant to PXO99 (Table 2). The result was not consistent with the previous finding that the *Xa21*gene in either IR24 or TP309 genetic background conferred complete resistance to PXO99 (Gu et al.[2004]; Ikeda et al. [1990]; Song et al. [1995]; Wang et al. [1996]).

Table 2: Disease evaluation of WH421, Siputeh and TS4 for resistance to*Xoo*strains

		Lesion length (cm) and disease phenotype[a]		
Stains	**Origin**	**WH421**	**Siputeh**	**TS4**
1947	Africa	14.8 ± 7.4 (S)	24.1 ± 6.9 (S)	4.4 ± 4.5 (MR)
Aust-2031	Australia	2.8 ± 1.3 (R)	25.5 ± 8.4 (S)	1.3 ± 0.5 (R)
Aust-R3	Australia	1.8 ± 1.1 (R)	39.5 ± 8.8 (S)	1.0 ± 0.5 (R)
GD1358	China	7.0 ± 1.7 (MS)	23.9 ± 4.7 (S)	4.8 ± 3.6 (MR)
HB17	China	0.5 ± 0.3 (R)	31.2 ± 3.8 (S)	0.9 ± 0.6 (R)
HB21	China	1.7 ± 0.9 (R)	39.6 ± 6.7 (S)	1.3 ± 0.4 (R)
HLJ72	China	1.5 ± 0.6 (R)	16.2 ± 3.3 (S)	0.7 ± 0.4 (R)
JS49-6	China	0.9 ± 0.5 (R)	13.6 ± 4.9 (S)	1.3 ± 0.6 (R)
LN57	China	0.7 ± 0.4 (R)	30.7 ± 8.5 (S)	1.1 ± 0.8 (R)
NX42	China	1.3 ± 0.8 (R)	29.7 ± 5.6 (S)	1.5 ± 1.0 (R)
ZHE173	China	0.2 ± 0.1 (R)	34.2 ± 7.8 (S)	1.1 ± 0.8 (R)
CIAT1185	Columbia	3.5 ± 2.0 (MR)	10.0 ± 3.5 (S)	2.3 ± 0.7 (R)
A3842	India	4.3 ± 1.2 (MR)	37.9 ± 13.2 (S)	3.0 ± 1.1 (R)

A3857	India	18.8±4.9 (S)	38.3±7.4 (S)	10.5±2.4 (S)
IXO56	Indonesia	6.0±2.2 (MR)	30.3±6.7 (S)	3.7±2.0 (MR)
H75373	Japan	0.6±0.6 (R)	25.5±4.9 (S)	0.4±0.1 (R)
T7174	Japan	0.6±0.6 (R)	29.9±6.7 (S)	0.6±0.5 (R)
JW89011	Korea	16.1±8.9 (S)	35.9±7.0 (S)	9.2±4.3 (S)
NXO260	Nepal	3.8±6.1 (MR)	22.1±2.6 (S)	1.1±1.0 (R)
K202	Korea	15.1±3.7 (S)	44.9±10.7 (S)	17.0±4.3 (S)
PXO71	Philippines	0.4±0.3 (R)	37.8±14.6 (S)	1.1±0.7 (R)
PXO79	Philippines	0.7±0.5 (R)	20.4±7.5 (S)	2.8±1.5 (R)
PXO86	Philippines	1.9±0.8 (R)	28.7±9.6 (S)	2.8±1.2 (R)
PXO99	Philippines	11.8±2.8 (S)	25.0±7.1 (S)	5.9±3.0 (MR)
PXO112	Philippines	0.6±0.9 (R)	10.8±3.3 (S)	1.1±2.3 (R)
PXO113	Philippines	0.6±0.4 (R)	27.8±7.8 (S)	0.9±0.6 (R)
Thai R7	Thailand	1.3±0.5 (R)	13.2±6.0 (S)	0.9±0.7 (R)
Thai R2	Thailand	2.8±1.3 (R)	35.4±5.6 (S)	2.2±1.4 (R)

[a]R, resistant (Lesion length ≤3.0 cm); MR, moderately resistant (3.0 cm < Lesion length ≤6.0 cm); MS, moderately susceptible (6.0 cm < Lesion length ≤9.0 cm); S, susceptible (Lesion length >9.0 cm).

Luo *et al.*

Luo *et al. Rice* **2014 7:33** **doi**:10.1186/s12284-014-0033-2

Field Evaluation of TS4 and Siputeh

The agronomic traits of TS4 and Siputeh were evaluated in two field trials and different growth seasons conducted in Banda Aceh, Indonesia and Lingshui, China, respectively. Due to presence of the *sd1* gene, TS4 (123.7 ± 2.1 cm) had shorter plant height than Siputeh (208.3 ± 3.9 cm) when they were grown in Banda Aceh (Table 3). The difference in plant height between Siputeh (214.6 ± 4.4 cm) and TS4 (106.7 ± 1.1 cm) became even greater when they were grown in Lingshui under short-day condition during the winter season (Figure 3). TS4 (117 days) had shorter growth duration than Siputeh (160 days) when they were planted in Banda Aceh, Indonesia (Table 3). However, both TS4 (162 days) and Siputeh (182 days) had long growth duration when they were planted in Lingshui under short-day condition during the winter (Table 3). TS4 produced more productive panicles per plant with higher seed-setting rate but fewer grain number per panicle than Siputeh grown in both field trials (Table 3). The greater number of productive panicles per plant and higher seed-setting rate also collectively contributed to the higher grain yield of TS4 than that of Siputeh grown in both field trails (Table 3). TS4 (27.6 g) had similar 1000-grain weight to Siputeh (27.4 g) when they were grown in Banda Aceh (Table 3). However, the 1000-grain weight of TS4 (34.1 g) were slightly heavier than that of Siputeh (31.7 g) when they were grown in Lingshui (Table 3). In summary, TS4 showed better agronomic traits with shorter plant height and growth duration and higher yield than Siputeh.

Table 3: Agronomic traits of Siputeh and TS4 plants grown in field trials

	Field trial 1[a]			**Field trial 2**[b]		
Agronomic trait	**Siputeh**	**TS4**	***t*-testc**	**Siputeh**	**TS4**	***t*-test**
Growth duration (days)	160	117	-	182	162	-
Plant height (cm)	208.3 ± 3.9	123.7 ± 2.1**	33.081	214.6 ± 4.4	106.7 ± 1.1**	41.120
Productive panicles per plant	6.5 ± 0.8	9.0 ± 0.4**	4.841	4.8 ± 0.2	5.7 ± 0.5*	3.298
Grain number per panicle	137.3 ± 3.4	129.7 ± 2.9*	2.946	172.4 ± 11.5	139.0 ± 16.3*	2.902
Seed-setting rate (%)	70.1 ± 2.8	80.6 ± 2.6**	4.760	70.5 ± 1.2	82.8 ± 2.0**	9.097
1000-grain weight (g)	27.4 ± 0.7	27.6 ± 0.4	0.430	31.7 ± 0.6	34.1 ± 0.5**	8.978
Yield (t/ha)	4.3 ± 0.3	6.5 ± 0.1**	12.050	6.6 ± 0.7	8.0 ± 0.2*	3.538

[a]Field trial 1 was conducted in Banda Aceh, Indonesia, in the wet season of 2011/2012 (November 2011 to April 2012).

[b]Field trial 2 was conducted in Lingshui, China, in the winter season of 2013/2014 (November 2013 to May 2013).

[c]*t*-test is the comparision between Siputeh and TS4. $T_{0.05,\,4} = 2.776$, $T_{0.01,\,4} = 4.604$.

"**" and "*" stand for significance difference at 0.01 and 0.05 probability levels, respectively.

Luo *et al.*

Luo *et al.* *Rice* 2014 **7**:33 doi:10.1186/s12284-014-0033-2

Plant height and maturation time of Siputeh and TS4 in the field. The semi-dwarf TS4 plants matured earlier than Siputeh with tall plant type. Picture was taken in the field in Lingshui, China, at 162 days after sowing.

Luo *et al.* *Rice* 2014 **7**:33 doi:10.1186/s12284-014-0033-2

Comparison of Grain Quality between TS4 and Siputeh

The grain quality of TS4 and Siputeh was measured using rice grains harvested from the two field trials. Both TS4 and Siputeh produce long grain rice with similar grain length and length-to-width ratio (Table 4). TS4 also has significantly lower degree of chalkiness than Siputeh in both field trials (Table 4). TS4 has lower amylose content (15.3-17.8%) than Siputeh (26.3-28.9%), due to the presence of the Wxb gene in TS4 (Table 4). For the rice grains harvested in field trial in Banda Aceh, TS4 has soft gel consistency while Siputeh had medium gel consistency (Table 4). Both TS4 and Siputeh had intermediate gelatinization temperature even though TS4 had slightly lower akali spreading value than Siputeh (Table 4). The results collectively showed that TS4 has better grain quality than Siputeh in terms of degree of chalkiness and amylose content.

Table 4: Grain quality of Siputeh and TS4

	Field trial 1[a]			Field trail 2[b]		
Traits	**Siputeh**	**TS4**	***t*-test[c]**	**Siputeh**	**TS4**	***t*-test**
Grain length (mm)[d]	7.4±0.1 (Long)	7.6±0.1 (Long)	2.449	7.2±0.1 (Long)	7.3±0.1 (Long)	1.225
Length-to-width ratio[e]	3.2±0.1 (Slender)	3.1±0.1 (Slender)	1.225	3.4±0.1 (Slender)	3.2±0.1 (Slender)	2.449
Degree of chalkiness (%)[f]	23.0±2.0 (9)	12.0±1.0** (5)	18.249	16.3±0.6 (5)	9.3±0.3** (1)	18.074
Amylose content (%)[g]	26.3±0.1 (High)	15.3±0.2** (Low)	67.361	28.9±0.2 (High)	17.8±0.2** (Low)	67.973
Gel consistency (mm)[h]	51.0±1.6 (Medium)	90.0±2.7** (Soft)	23.735	68.8±5.0 (Soft)	94.5±0.9** (Soft)	8.762
Alkali spreading value and gelatinization temperature[i]	5.0±0.1 (Intermediate)	4.5±0.1** (Intermediate)	8.660	5.0±0.0 (Intermediate)	4.0±0.1** (Intermediate)	6.124

[a]Field trial 1 was conducted in Banda Aceh, Indonesia, in the wet season of 2011/2012 (November 2011 to April 2012).

[b]Field trial 2 was conducted in Lingshui, China, in the winter season of 2013/2014 (November 2013 to May 2013).

[c]*t*-test is the comparision between Siputeh and TS4. $T_{0.05,\ 4}$ =2.776, $T_{0.01,\ 4}$ =4.604. "**" and "*" stand for significance difference at 0.01 and 0.05 probability levels, respectively.

[d]Category of grain length: Very long, grain length>7.5 mm; Long, 6.6 mm <grain length≤7.5 mm; Medium, 5.5 mm <grain length≤6.6 mm; Short, grain length≤5.5 mm.

[e]Grain shape based on length-to-width ratio: Slender, length-to-width ratio>3.0; Medium, 2.0 <length-to-width ratio≤3.0; Bold, length-to-width ratio≤2.0.

[f]Scale for degree of chalkiness: 0, degree of chalkiness =0; 1, 0 <degree of chalkiness≤10%; 5, 10%<degree of chalkiness≤20%, 9, degree of chalkiness>20%.

[g]Classification of amylose content: Waxy, amylose content≤2%; Very low, 2% <amylose content≤9%; Low, 9% <amylose content≤20%; Intermediate, 20% <amylose content≤25%; High, amylose content>25%.

[h]Classification of gel consistency: Soft, gel consistency>60 mm; Medium, 40 mm <gel consistency≤60 mm; Hard, gel consistency≤40 mm.

[i]Grade of gelatinization temperature estimated by alkali spreading value: High (74.5°C≤gelatinization temperature <80°C), 1≤alkali spreading value <2.5; Intermediate high (74°C≤gelatinization temperature <74.5°C), 2.5≤alkali spreading

value <3.5; Intermediate (70°C≤gelatinization temperature <74°C), 3.5≤alkali spreading value <5.5; Low (gelatinization temperature <70°C), 5.5≤alkali spreading value≤7.

Luo *et al.*

Luo *et al. Rice* 2014 **7**:33 doi:10.1186/s12284-014-0033-2

DISCUSSION

Marker-assisted selection is a molecular breeding process whereby a molecular marker based on DNA variation is used for indirect selection of an interest agronomic trait. Marker-assisted selection is not affected by environments and can be conducted at any plant growth stage. It is especially useful for selection of recessive genes and biochemical traits in heterozygous plants. In this study, the *sd1* gene is a partially recessive gene that controls semi-dwarf phenotype. It is difficult or impractical to select the *sd1* gene in the backcrossing progeny according to plant height, especially under greenhouse condition. Similarly, the amylose content in rice grains can only be directly detected after harvesting, however, we need to know the genotype for crossing purpose before flowering. As for the pyramiding of *Xa4* and *Xa21* for bacterial blight resistance, both *R* genes have different resistance specificities and their resistance spectrums are over-lapped. It would be impractical to select the two *R* genes in a single line by conducting disease evaluation and phenotypic selection. To overcome these difficulties, we chose marker-assisted selection approach, which enable us to perform precisely and effectively multiple gene selection and/or pyramiding in limited breeding time.

Compared with Siputeh, the significant change in TS4 is the semi-dwarf plant type, which is mainly resulted from the introduction of the *sd1* gene. The semid-warf plant type of rice indiredtly contributes to lodging resistance in the field. It also increases grain-straw ratio, which ultimately results in higher yield. In the present study, TS4 produced higher yields than Siputeh in both field trials conducted in different location (Table 3). The results are consistent with the results of previously studies on the *sd1* gene introduced in other varieties (Hedden [2003]; Luo and Yin[2013]; Spielmeyer et al. [2002]). TS4 also showed significant reduction in growth duration than Siputeh. The growth duration of TS4 was about four months when it was grown in

Banda Aceh, Indonesia (Table 3), which was similar to most of the typical modern rice varieties grown in tropical region. Short-growth duration makes it easier for farmers to plan growth season, synchronize sowing and harvest time with other varieties and manage diseases and insects in the field.

Like the *Xa4* and *Xa21* donor line WH421, TS4 provided broad-spectrum resistance to multiple *Xoo*strains tested in this study. It was noted that WH421 were susceptible or moderate susceptible to 1947, PXO99 and GD1358 while TS4 provided moderate resistance to the three *Xoo* strains (Table 2). It seems that either *Xa4* or *Xa21* or both *R* genes performed better for resistance to *Xoo*in TS4 than that in WH421. Previously, it was found that the *Xa21* gene in IRBB21 conferred complete resistance to PXO99 (Gu et al. [2004]; Ikeda et al. [1990]; Song et al. [1995]), however, in this study, neither WH421 nor TS4 could confer full resistance to the *Xoo* strain. It has been reported that the *R* gene-mediated resistance to *Xoo* in rice may be influenced by rice genetic backgrounds and developmental stages (Cao et al. [2007]). For instance, the *Xa21*-containing rice line Minghui was still susceptible to PXO99 (Zhang et al. [2006]).

The introduction of the *sd1* gene to Siputeh would significantly reduce its plant height and growth duration, which might affect rice grain quality. To overcome this potential shortcoming, we also introduced the Wx^b gene into TS4 through marker-assisted selection. Indeed, the Wx^b gene significantly reduced the amylose content in TS4. Amylose content strongly affects the cooking and eating quality of rice. Rice with amylose content at 25-30% usually tends to be firm and dry after cooking, rice with amylose content at 20-25% tends to be softer and stickier and rice with amylose content at <20% is generally quite soft and sticky. TS4 has amylose content at 15.3-17.8% and shows soft gel consistency. The results indicate that TS4 is generally softer and stickier than Siputeh, which has amylose content at 26.3-28.9% and shows medium to soft gel consistency.

CONCLUSION

An improved rice line, designed as TS4, has been developed to contain semi-dwarf gene *sd1*, low amylase content gene Wx^b and bacterial light resistance genes *Xa4* and *Xa21* through marker-assisted selection. TS4 has semi-dwarf phenotype with reduced growth duration, produces high yield with good grain quality and provides broad-spectrum resistance to *Xoo* strains. The development of TS4 enriches the diversity of local rice varieties with high yield potential and good grain quality.

METHODS

Plant Materials and Growth Condition

Rice variety Siputeh is a local cultivar collected in Aceh province, Indonesia. Rice variety WH421 carrying *sd1*, Wx^b, *Xa4* and *Xa21* genes is a paternal line of hybrid rice in Mianhui 725 backgroud (Luo et al. [2012]). Rice plants were grown in the greenhouse at 32°C for 12.5 h (light) and 25°C for 11.5 h (dark).

PCR-Based Molecular Markers

The molecular markers used in this study include the codominant STS (Sequence-Tagged Sites) marker sd1STS for the *sd1* gene (Luo and Yin [2013]), the co-dominant microsatellite DNA marker RM224 for the *Xa4* gene (Sun et al. [2003]), the co-dominant STS marker 21 for the *Xa21* gene (Chen et al. [2000]) and the co-dominant CAPS (Cleaved Amplified Polymorphic Sequences) marker PCR-*AccI* for the Wx^a and Wx^b genes (Cai et al. [2002]). Oligo primers for PCR amplification of the molecular markers are listed in Table 1.

PCR Amplification

PCR was performed on a PTC-100 programmable thermal controller (MJ Research). The PCR reaction mixture of 20 μl consisted of 1–100 ng of rice genomic DNA, 0.2 mM each of dNTPs, 0.2 μM of each primer, 2 μl of 10 × PCR buffer [500 mM KCl, 100 mM Tris–HCl (pH 8.3), 20 mM $MgCl_2$] and 0.5 unit of Taq polymerase. For PCR amplification of markers sd1STS, 4 ul of 5 × Q-solution (QIAGEN) was added to the PCR reaction mixture. Template DNA was initially denatured at 94°C for 2 min followed by 35 cycles of PCR amplification with the following parameters: a 30 s of denaturation at 94°C, a 40 s of primer annealing at 55°C for markers 21, RM224 and PCR-*Acc*I, 65°C for maker sd1STS, and 1 min of primer extension at 72°C for markers RM224 and PCR-*Acc*I, and 1.5 min for markers 21 and sd1STS. Finally, the reaction mixture was maintained at 72°C for 5 min of primer extension before completion. The PCR product of the marker PCR-*Acc*I was digested by *Acc*I for 4 h. The amplified products were electrophoretically resolved on a 1.5% agarose gel for marker sd1STS, a 2.0% agarose gel for marker 21, and a 3.5% agarose gel for marker PCR-*Acc*I and RM224 in 1 × TAE buffer.

Bacterial Inoculation and Disease Scoring

Xoo strains were grown on PSA medium (10 g/l peptone, 10 g/l sucrose, 1 g/lglutamic acid, 16 g/l bacto-agar, and pH7.0) for about 60 hours at 28°C. Bacterial cells were suspended in sterile water and diluted to an opticla density (OD) at 600 nm of 0.5. Plants were inoculated by the leaf-clipping method (Kauffman et al. [1973]). Lesion length (L.L.) was measured at 14 days after inoculation. The disease symptoms were scored as resistant (R, L.L. ≤ 3.0 cm), moderately resistant (MR, 3.0 cm < L.L. ≤ 6.0 cm), moderately susceptible (MS, 6.0 cm < L.L. ≤ 9.0 cm) or susceptible (S, L.L. > 9 cm).

Field Trial Design and Collection of Important Agronomic Traits

Field trials were conducted in Banda Aceh, Indonesia in the season of 2011/2012 (November 2011 - March 2012) and in Lingshui, China in the winter season of 2013/2014 (November 2013 - May 2014), respectively. Rice lines were tested in replicated plots in each field trial. The plot size was 4.2 m × 3.8 m and the plant spacing was 20 cm × 20 cm in Banda Aceh, while the plot size was 4 m × 4 m and the plant spacing was 16.7 cm × 16.7 cm in Lingshui. Thirty plants (3 × 10 plants/per plot) were randomly selected for each tested line and scored for important agronomic traits, including growth duration, plant height, productive panicle per plant, number of grains per panicle, seed setting rate, weight of 1000 grains and grain yield per hectare. Statistical analysis was performed using a two-tailed *t*-test for independent samples.

Evaluation of Grain Quality

Rice grain quality parameters, including grain length, the ratio of length to width (L/W), chalkiness degree, amylose content (AC), gel consistency (GC) and alkali spreading value (ASV) were measured by the methods described previously (Cruz and Khush [2000]).

Abbreviations

CAPS: Cleaved amplified polymorphic sequence

DNA: Deoxyribonucleic acid

GBSSI: Granule-bound starch synthase I

MR: Moderately resistant

MS: Moderately susceptible

OD: Opticla density

PCR: Polymerase Chain Reaction

R: Resistant

S: Susceptible

SSR: Simple sequence repeat

STS: Sequence-tagged site

Xoo: *Xanthomonas oryzae* pv. *oryzae*

Competing Interests

The authors declare that they have no competing interests.

Authors' Contribution

YL and ZY designed the experiments. YL conducted the experiments with the help from SZ, BB, TM, ZL and JY. YL and ZY wrote the manuscript. All authors read and approved the final manuscript.

AUTHORS' INFORMATION

YL is the Senior Research Officer of Temasek Life Science Laboratory (TLL) and the Professor of Anhui Rice Research Institute (ARRI), Anhui Academy of Agricultural Sciences (AAAS), with over 25 years of experiences on rice breeding; SZ is the Professor of Agriculture Faculty, Syiah Kuala University (SKU); BB is the Associate Professor of Agriculture Faculty, SKU; TM is the Associate Professor of ARRI, AAAS; ZL is the Director and Professor of ARRI, AAAS; JY is the Professor of ARRI and the President of AAAS; ZY is the Senior Principal Investigator of TLL.

ACKNOWLEDGEMENTS

The authors wish to thank Mr Y. Zhou for technical assistance in field trials. This research is supported by a grant from Temasek Foundation to Z. Yin.

REFERENCES

1. Asano K, Takashi T, Miura K, Qian Q, Kitano H, Matsuoka M, Ashikari M (2007) Genetic and molecular analysis of utility of *sd1* alleles in rice breeding. Breed Sci57:53-58
2. Ashikari M, Sasaki A, Ueguchi-Tanaka M, Itoh H, Nishimura A, Datta S, Ishiyama K, Saito T, Kobayashi M, Khush GS (2002) Loss-of-function of a rice gibberellin biosynthetic gene, GA20 oxidase (GA20ox-2), led to the rice'green revolution'. Breed Sci 52:143-150
3. Cai XL, Wang ZY, Xing YY, Zhang JL, Hong MM (1998) Aberrant splicing of intron 1 leads to the heterogeneous 5 UTR and decreased expression of waxy gene in rice cultivars of intermediate amylose content. Plant J 14:459-465 doi: 10.1046/j.1365-313X.1998.00126.x
4. Cai XL, Liu QQ, Tang SZ, Cu MH, Wang ZY (2002) Development of molecular marker for screening the rice cultivars with intermediate amylose content in *Oryza sativa* subsp. *indica*. J

Plant Physiol and Mol Biol 28:137-144

5. Cao YL, Ding XH, Cai M, Zhao J, Lin YJ, Li XH, Xu CG, Wang SP (2007) Expression pattern of a rice disease resistance gene *Xa3/ Xa26* is differentially regulated by the genetic backgrounds and developmental stages that influence its function. Genetics 177:523-533, doi:10.1534/genetics.107.075
6. Chen S, Lin X, Xu C, Zhang Q (2000) Improvement of bacterial blight resistance of 'Minghui 63', an elite restorer line of hybrid rice, by molecular marker-assisted selection. Crop Sci 40:239-244
7. Cruz ND, Khush G (2000) Rice Grain Quality Evaluation Procedures. In: Singh RK, Singh US, Khush GS (eds) Aromatic Rices, Oxford and IBH Pulishing Co. Pvt. Ltd, New Dehli, Calcutta. pp 24-36
8. Datta K, Baisakh N, Thet KM, Tu J, Datta SK (2002) Pyramiding transgenes for multiple resistance in rice against bacterial blight, yellow stem borer and sheath blight. Theor Appl Genet 106:1-8
9. Gnanamanickam SS, Priyadarisini VB, Narayanan NN, Vasudevan P, Kavitha S (1999) An overview of bacterial blight disease of rice and strategies for its management. Curr Sci India 77:1435-1444
10. Gu K, Tian D, Yang F, Wu L, Sreekala C, Wang D, Wang GL, Yin Z (2004) High-resolution genetic mapping of *Xa27* (t), a new bacterial blight resistance gene in rice, *Oryza sativa* L. Theor Appl Genet 108:800-807
11. Hedden P (2003) The genes of the green revolution. Trends Genet 19:5-9 doi:S0168-9525(02)00009-4
12. Hori Y, Fujimoto R, Sato Y, Nishio T (2007) A novel wx mutation caused by insertion of a retrotransposon-like sequence in a glutinous cultivar of rice (*Oryza sativa*). Theor Appl Genet 115:217-224
13. Huang N, Angeles ER, Domingo J, Magpantay G, Singh S, Zhang G, Kumaravadivel N, Bennett J, Khush GS (1997) Pyramiding of bacterial blight resistance genes in rice: marker-assisted selection using RFLP and PCR. Theor Appl Genet 95:313-320
14. Ikeda R, Khush G, Tabien R (1990) A new resistance gene to bacterial blight derived from *O. longistaminata*. Jpn J Breed 40:280-281

15. Jennings PR (1964) Plant type as a rice breeding objective. Crop Sci 4:13-15
16. Kauffman H, Reddy A, Hsieh S, Merca S (1973) An improved technique for evaluating resistance of rice varieties to *Xanthomonas oryzae*. Plant Dis Rep 57:537-541
17. Larkin PD, Park WD (1999) Transcript accumulation and utilization of alternate and non-consensus splice sites in rice granule-bound starch synthase are temperature-sensitive and controlled by a single-nucleotide polymorphism. Plant Mol Biol 40:719-727
18. Luo Y, Yin Z (2013) Marker-assisted breeding of Thai fragrance rice for semi-dwarf phenotype, submergence tolerance and disease resistance to rice blast and bacterial blight. Mol Breed 32:709-721, doi:10.1007/s11032-013-9904-2
19. Luo Y, Sangha JS, Wang S, Li Z, Yang J, Yin Z (2012) Marker-assisted breeding of Xa4, Xa21 and Xa27 in the restorer lines of hybrid rice for broad-spectrum and enhanced disease resistance to bacterial blight. Mol Breed 30:1601-1610
20. Mew TW, Alvarez AM, Leach JE, Swings J (1993) Focus on bacterial blight of rice. Plant Dis 77:5-12
21. Sano Y (1984) Differential regulation of Waxy gene expression in rice endosperm. Theor Appl Genet 68:467-473
22. Sasaki A, Ashikari M, Ueguchi-Tanaka M, Itoh H, Nishimura A, Swapan D, Ishiyama K, Saito T, Kobayashi M, Khush G (2002) A mutant gibberellin-synthesis gene in rice. Nature 416:701-702
23. Singh S, Sidhu JS, Huang N, Vikal Y, Li Z, Brar DS, Dhaliwal HS, Khush GS (2001) Pyramiding three bacterial blight resistance genes (*xa5*, *xa13* and *Xa21*) using marker-assisted selection into indica rice cultivar PR106. Theor Appl Genet 102:1011-1015
24. Song W-Y, Wang G-L, Chen L-L, Kim H-S, Pi L-Y, Holsten T, Gardner J, Wang B, Zhai W-X, Zhu L-H, Fauquet C, Ronald PC (1995) A receptor kinase-like protein encoded by the rice disease resistance gene, Xa21. Science 270:1804-1806
25. Spielmeyer W, Ellis MH, Chandler PM (2002) Semidwarf (*sd-1*), "green revolution" rice, contains a defective gibberellin 20-oxidase gene. Proc Natl Acad Sci USA 99:9043-9048
26. Su Y, Rao Y, Hu S, Yang Y, Gao Z, Zhang G, Liu J, Hu J, Yan M,

Dong G (2011) Map-based cloning proves qGC-6, a major QTL for gel consistency of japonica/indica cross, responds by Waxy in rice (*Oryza sativa* L.). Theor Appl Genet 123:859-867

27. Suh JP, Jeung JU, Noh TH, Cho YC, Park SH, Park HS, Shin MS, Kim CK, Jena KK (2013) Development of breeding lines with three pyramided resistance genes that confer broad-spectrum bacterial blight resistance and their molecular analysis in rice. Rice 6:1-11
28. Suhartini T (1991) Preliminary Evaluation of Performances and Yields of Rice Varieties and Lines at Lebak area. Paper presented at the Proceedings of a seminar of research outcomes of Bogor Research Institute of food crops, Bogor (Indonesia), 11–12 March, 1991
29. Sun X, Yang Z, Wang S, Zhang Q (2003) Identification of a 47-kb DNA fragment containing *Xa4*, a locus for bacterial blight resistance in rice. Theor Appl Genet 106:683-687
30. Tian Z, Qian Q, Liu Q, Yan M, Liu X, Yan C, Liu G, Gao Z, Tang S, Zeng D (2009) Allelic diversities in rice starch biosynthesis lead to a diverse array of rice eating and cooking qualities. Proc Natl Acad of Sci USA 106:21760-21765
31. Wang ZY, Zheng FQ, Shen GZ, Gao JP, Snustad DP, Li MG, Zhang JL, Hong MM (1995) The amylose content in rice endosperm is related to the posttranscriptional regulation of the waxy gene. Plant J 7:613-622
32. Wang G-L, Song W-Y, Ruan D-L, Sideris S, Ronald PC (1996) The cloned gene, Xa21, confers resistance to multiple *Xanthomonas oryzae* pv. *oryzae* isolates in transgenic plants. Mol Plant-Microbe Interact 9:850-855
33. Zhang Q (2009) Genetics and improvement of bacterial blight resistance of hybrid rice in China. Rice Sci 16:83-92
34. Zhang J, Li X, Jiang G, Xu Y, He Y (2006) Pyramiding of *Xa7* and *Xa21* for the improvement of disease resistance to bacterial blight in hybrid rice. Plant Breed 125:600-605
35. Zhu C, Shen W, Zhao H, Wan J (2003) Advances in researches of the application of low-amylose content rice gene for breeding. Sci Agri Sinica 37:157-162

Chapter 2

The Roots of Future Rice Harvests

Nourollah Ahmadi[1], Alain Audebert1, Malcolm J Bennett[2], Anthony Bishopp[2], Antonio Costa de Oliveira[3], Brigitte Courtois[1], Abdala Diedhiou[45], Anne Diévart[1], Pascal Gantet[67], Alain Ghesquière[8], Emmanuel Guiderdoni[1*], Amelia Henry[9], Yoshiaki Inukai[10], Leon Kochian[11], Laurent Laplaze[58], Mikael Lucas[8], Doan Trung Luu[12],Baboucarr Manneh[13], Xiaorong Mo[14], Raveendran Muthurajan[15], Christophe Périn[1],Adam Price[16], Sabariappan Robin[15], Hervé Sentenac[12], Bassirou Sine[175], Yusaku Uga[18], Anne Aliénor **Véry**[12], Matthias Wissuwa[19], Ping Wu14 and Jian Xu[20]

[1]CIRAD, UMR AGAP, Montpellier Cedex 5, 34398, France

[2]Centre for Plant Integrative Biology, University of Nottingham, Loughborough LE12 5RD, UK

[3]UFPel, Plant Genomics and Breeding Center, Pelotas, Brazil

[4]Université Cheikh Anta Diop (UCAD), Département de Biologie Végétale, Laboratoire Commun de Microbiologie IRD/ISRA/UCAD, Centre de

Recherche de Bel Air - BP 1386 CP 18524, Dakar, Sénégal

[5]Laboratoire Mixte International Adaptation des Plantes et microorganismes associés aux Stress Environnementaux (LAPSE), Centre de Recherche de Bel Air - BP 1386 CP 18524, Dakar, Sénégal

[6]Université Montpellier 2, UMR DIADE, Montpellier, France

[7]IRD, LMI RICE, USTH, Agronomical Genetics Institute, Hanoi, Vietnam

[8]IRD, UMR DIADE, Montpellier, France

[9]IRRI, Los Baños, Philippines

[10]International Cooperation Center for Agricultural Education (ICCAE), Nagoya University, Furo-cho, Chikusa 464-8601, Nagoya, Japan

[11]Robert W. Holley Center for Agriculture and Health, USDA-ARS and Department of Plant Biology, Cornell University, Ithaca 14853, NY, USA

[12]Biochimie et Physiologie Moléculaire des Plantes, Institut de Biologie Intégrative des Plantes, UMR 5004 CNRS/386 INRA/Montpellier SupAgro/ Université Montpellier 2, F-34060, Montpellier Cedex 2, France

[13]Africa Rice Center, AfricaRice Sahel Regional Station, B.P. 96, St Louis, Senegal

[14]State Key Laboratory of Plant Physiology and Biochemistry, College of Life Science, Zhejiang University, Hangzhou 310058, China

[15]Tamil Nadu Agricultural University, Coimbatore, 641003, India

[16]University of Aberdeen, Aberdeen AB24 3UU, UK

[17]ISRA, CERAAS, Thiès, Senegal

[18]National Institute of Agrobiological Sciences (NIAS), 2-1-2 Kannondai, Tsukuba 305-8602, Ibaraki, Japan

[19]Japan International Research Center for Agricultural Sciences (JIRCAS), 1-1 Ohwashi, Tsukuba 305-8686, Japan

[20]Department of Biological Sciences and NUS Centre for BioImaging Sciences, Faculty of Science, National University of Singapore, Singapore 117543, Singapore

ABSTRACT

Rice production faces the challenge to be enhanced by 50% by year 2030 to meet the growth of the population in rice-eating countries. Whereas yield of cereal crops tend to reach plateaus and a yield is likely to be deeply affected by climate instability and resource scarcity in the coming decades, building rice cultivars harboring root systems that can maintain performance by capturing water and nutrient resources unevenly distributed is a major breeding target. Taking advantage of gathering a community of rice root biologists in a Global Rice Science Partnership workshop held in Montpellier, France, we present here the recent progresses accomplished in this area and focal points where an international network of laboratories should direct their efforts.

REVIEW

The root system has the crucial role of extracting nutrients and water through a complex interplay with soil biogeochemical properties, and of maintaining these functions under a wide range of stress scenarios to ensure plant survival and reproduction. This role is made even more important due to increasing climate instability and limitation of fertilizers and irrigation in cropping systems. Determining the precise contribution of root traits to final grain yield under these scenarios and breeding cultivars harboring root systems adapted to stress profiles prevalent in representative target soil environments are therefore a priority in the plant breeding research agenda. Long neglected, the biology and ecology of roots - the "hidden half" - have recently attracted an increasing number of research groups and disciplines ranging from genetics, molecular biology, cell biology, physiology, microbiology, engineering, and biomathematics. This attraction has been exemplified by the high attendance of the International Root Research Society meetings (http://www.rootresearch.org *webcite*). Demonstrations at the field level that root anatomical and architectural traits can considerably enhance cereal crop yield stability under limited water and nutrient resources (Lopes and Reynolds [2010], York et al. [2013], Lynch et al. [2014]), as well as the development of high-throughput root phenotyping systems (Clark et al. [2013], Topp et al.[2013]) have also recently contributed to this interest. Furthermore, more groups working in root development in the model plant *Arabidopsis* are now turning to crops, such as rice (*Oryza sativa*L.), the model cereal.

Addressing issues of water and nutrient limitations in rice, the staple food for more than half of mankind, is of particular relevance since rice is cultivated under the widest range of agro-systems among cereals - from irrigated anaerobic conditions to upland aerobic conditions, including flood-prone and drought-prone environments. Breeding for high-yielding irrigated rice has favored shallow root system phenotypes that easily capture resources in the topsoil layer whereas breeding for low input upland conditions has, conversely, favored deep root systems with strong foraging capacity and nutrient extraction ability from poor soil layers. In addition to these divergent root architectures that have been selected for among rice agro-ecosystems, the increasingly fluctuating levels of stress that is projected to occur across rice growing systems may necessitate more diverse root architectures and better adaptability to uneven distribution of soil resources. The diversity of rice for adaptation to different water regimes and soil conditions has favored the identification of genes and quantitative trait loci (QTLs) underlying root development and resource uptake traits. In parallel, methodological progress in the development of experimental set ups that may mimic uneven resource distribution, has led to the establishment of diverse and complementary phenotyping platforms for screening for root traits in diversity panels, mapping populations and mutant resources. Though their relevance to field soil conditions has to be considered with caution, such platforms are instrumental for feeding, then calibrating and validating, dynamic models of root architecture in response to resource scarcity and distribution (reviewed in Hill et al. [2013]). Modeling is likely to be of invaluable assistance for breeding ideotypes adapted to specific soil environments and stress scenarios as millions of simulations can be run encompassing a wide range of environmental conditions (Postma et al. [2014]).

To review the progress in these areas and delineate future priorities where research efforts on rice roots should focus, a workshop gathering 40 participants was organized in the Agropolis campus, Montpellier France, from May 15-16th 2014, in the framework of the Global Rice Science Partnership (GRiSP). The group has first focused its debates on advances in root development and ideotype breeding. As a major output of the discussions, we recommend research areas on which emphasis should be placed that constitute a priority set of core actions of a rice root research collaborative network.

Accelerate the Discovery of Genes Underlying Root Structural and Functional Traits

During the past two years, the rice research community has seen major breakthroughs in the identification of genes behind root response to stress; *DRO1*, a root depth QTL (Uga et al., [2013]) and *PUP1* a root QTL for phosphorus uptake (Gamuyao et al., [2012]) were cloned, and fine mapping of several other rice root/stress response QTLs is ongoing (*DRO2, DRO3, CHR9;* Y. Uga and N. Ahmadi et al. unpublished work). Recent advances in MetaQTL analyses and genome wide association mapping approaches (GWAS) can reduce QTL confidence intervals and can help to improve the precision of the number and position of detected QTLs (Courtois et al. [2009],[2013]). These approaches are time consuming, and shortening the timeframe for QTL characterization is clearly a priority for the future.

The most effective strategy for cloning these QTLs appears to be the development of Near-Isogenic Lines (NILs) by marker-assisted selection. These NILs are useful material for fine mapping and QTL cloning but also for breeding and for detailed physiological studies. The NIL strategy combines breeding, exploitation of genetic diversity and positional cloning (Motte et al.[2014]). The work around root architecture and phosphate uptake also illustrates the importance of identifying cultivars that have special behavior (e.g. highly efficient P uptake despite a poor root system) to include rare alleles that will be valuable for breeding. Rice CSSL (Chromosome Segment Substitution Lines) populations, are valuable/natural starting points for QTL mapping and subsequent QTL cloning. Alternatively, developing new populations, especially NILs targeting QTLs of large effect is still a necessary step for its positional cloning.

The rice root system displays a complex structure composed of several tissues and different root types (Rebouillat et al. [2009]). The function of an increasing range of genes involved in rice root development has been deciphered these last 5 years (reviewed in Coudert et al. [2010]; Orman-Ligeza et al. [2013]). Root stress responses are clearly influenced by their cell identity and coordination of cell-specific responses that are currently poorly understood. All of the root tissues, including root cap (Wang et al. [2014]), could develop anatomical and molecular adaptations related to their diverse roles in root growth and function under normal and stress conditions. Exploring root stress responses and adaptations at a cell or tissue specific level by characterization of gene networks involved in these processes is an expanding research field (Gifford et al.[2008]). The use of gene regulatory network inference tools to identify key regulators of root adaptive responses need also to be used in rice. Such an approach should yield candidate master regulators of root responses to stress and is complementary to genetic approaches. The identification of common cis-acting regulatory elements can aid to gather co-expressed genes under root development and stress responses (Pegoraro et al. [2013]).

The validation of QTLs and reverse genetics often involves use of *A. thaliana* data. There is a need to improve and speed up annotation transfer between *Arabidopsis* and rice. Identification of orthologs is probably the most straightforward way. Expression data or synteny can be also used to find the right ortholog(s) between species. Databases of orthologs already exist (Goodstein et al. [2012]; Rouard et al. [2011]) as well as new tools for ortholog prediction (Ostlund et al. [2010]; Schreiber and Sonnhammer, [2013]) but the challenge is to build a simple and functional platform for biologists and breeders.

Functional validation of individual genes associated with root development through knockout or gain-of-function mutants still holds merit. Since the multiple genes of roots are found to have spatial and temporal differences in their expression and function, the search for the trait differences, genes and their function need to be expanded with large set of mutant resources (Wei et al. [2013] for a review). This includes the development of large saturation mutagenized populations, virtually exploring any amino acid variation along the encoded proteins. New gene targeting technologies, mainly TALEN and CRISPR are revolutionizing functional analysis of genes and should be critical to accelerate the characterization of genes of agronomic interest (Gaj et al.[2013]; Miao et al. [2013]). The establishment of an International repository containing CRISPR directed against all rice genes to generate knock-out mutants would be a helpful resource for speeding up gene discovery and validation. Another possibility would be to produce CRISPR in pairs targeting 5′ and 3′ end of each rice gene. This resource would be useful for allele replacement in any rice gene and to validate any QTL effect by introducing only the causative mutation in other genetic backgrounds. Finally, the CRISPR technology allows targeting multiple genes at once in a single transformation event and so opens a simple way to test QTL/gene interactions.

Establish High Throughput Phenotyping Platforms with Relevance to Field Conditions

GWAS is promoted by the recent progresses in re-sequencing rice genetic resources (CAAS BGI IRRI, [2014]) and reverse and direct genetics approaches are facilitated by the creation of large and diverse mutant collections (Wei et al. [2013]). Dedicated high-throughput phenotyping methods coupled with automated data analysis systems are necessary to fully exploit these genetic resources to identify the genetic determinants of root (adaptive) architecture.

The choice of a phenotyping method always results from a compromise between relevance, precision, throughput, automation and cost to address a specific scientific question. Some key root QTLs have been discovered indirectly without the necessity to screen roots. This is the case of PUP1 that was discovered merely on the basis of the capacity of plants to maintain yield when grown in phosphorus-depleted soils. It is only *a posteriori* that *PSTOL1*, the isolated gene underlying *PUP1* QTL, proved to control an adaptive development response of the root system in low-phosphate soil condition promoting root branching (Gamuyao et al. [2012]). Other important root genetic determinants governing rice adaptation to soil conditions could be found by this indirect way, but it will be probably limited. Most studies performed in the last decades were based on direct observation of the root system *per se*. The use of hydroponics-, soil pots- or soil columns-based systems and the collection of traits such as maximum root length, root thickness, root number, mass of roots at different depths, or root to shoot ratio allowed the identification of several QTLs (Courtois et al. [2009]). Some of these traits have already been integrated into breeding programs, e.g. to increase water deficit tolerance by increasing deep rooting (Shen et al.[2007]). Another soil-based and indirect root screening system uses introduction of herbicide at a deep soil layer (Khalfaoui and Havard, [1993]). In this case the plant starts to die when its root system reaches the herbicide layer. This reflects the rapidity with which the root develops in depth which is a combination of root gravity perception (or root angle) and growth rate (Al-Shugeairy et

scientific community to exploit common genetic resources (sequenced genotypes, highly recombinant populations, mutant collections) and to share data such as raw images that could be exploited for different phenotypic traits by different groups. This could apply for early developmental stage with simple standardized culture systems to investigate constitutive and adaptive root development and also for root anatomy. A common repository for these phenotyping data would also pave the way for analyses of genotype x environment interactions, meta-analyses and comparison between different experiments.

Delineate Root Ideotypes Adapted to Diverse Target Soil Environments

For future crops, it is expected that multiple root characteristics should be improved, including those that confer improved water use efficiency and nutrient acquisition (Pennisi [2010]). In the case of rice, one strategy has been to identify genes controlling such interesting root traits, which relies on the direct identification of genetic regions conferring tolerance to a given stress using a genetic screen, GWAS, or QTL analyses (e.g. *PUP1*; Gamuyao et al. [2012]). An alternative is to try to define *a priori* ideotypes of more efficient root systems from our knowledge of soil sciences and plant physiology. Root ideotypes that have been suggested to improve rice performance under drought include increased root growth at depth (Nicou et al. [1970]; O'Toole and Chang [1978]) and coarse axial roots to increase the ability to penetrate hardpans (Hasegawa et al. [1985]). The hypothesis that deep-rooted rice would be more resistant to drought than shallow rooted ones prompted the identification of a gene (*DRO1*) controlling nodal root growth angle in rice, and NIL having the functional allele of *DRO1* showed improved yield under upland drought conditions (Uga et al. [2013]). Similarly, deep rooting has been hypothesized as a trait that could allow for exploitation of soil nutrients at deeper soil layers thus improving mineral uptake and potentially increase yields (Obara et al. [2010]). NILs having the Kasalath allele at QTL qRL6.1 had longer seminal roots than those with the Koshihikari allele under different nitrogen fertilizer concentrations with NH_4^+ as the source of nitrogen. These examples demonstrate the potential value of ideotype-based approaches. However, as our knowledge of yield determination and response to stress increases, the complexity of the mechanisms involved in rice root response to stress and the diversity of environmental conditions in rice agro-ecosystems make the definition of ideotypes more difficult. Both the approaches of root ideotype-based screening and selection for grain yield are not mutually exclusive and should be conducted in parallel. They would in fact sustain each other; ideotype-based approaches may improve stress response and allow the testing of the ideotypes, and yield-based screens will provide better knowledge of the physiology of stress tolerance that will help refine the ideotypes. Such refined ideotypes should lead our formation of strategies to design new genetic screens based on a better knowledge of the integrated stress responses (physiology, development and plant-microbe interactions) and might help identify interesting alleles that might not be identified otherwise. For example, an experimental approach based on better knowledge of the physiology of water transport/acquisition has the potential to provide more detailed insights to mechanisms behind drought tolerance than a screen based on grain yield alone. Two major factors need to be addressed in order to optimize the benefit of such an ideotype-based approach: the understanding of environmental conditions and the integration of the complex factors behind stress response.

To optimize the outputs of root biology for crop improvement, it is first important to better define the environmental stresses that are faced by rice culture worldwide. This points to the need for better characterization of the actual (on-farm) target environments in which crops are grown (including soil physical properties, as in the approaches reported by Inthavong et al. [2011] or Haefele et al. [2014]), and the need for good estimates of future scenarios for environmental conditions faced by rice growers. This will help define the major limiting factors that can be addressed through root ideotype breeding, and to target the most important varieties specific to target each type of environmental conditions to maintain food production in the future (i.e., upland and rainfed lowland drought stress, or fluctuating soil moisture conditions).

Plant responses to environmental stresses are complex and involve multiple levels of integration. Root architecture has been the focus of most proposed ideotypes in the past but anatomical and physiological traits are likely to play equally key roles. Some research on water transport ideotypes has been conducted through studies on the effect of xylem vessel diameter (Yambao et al. [1992]; Henry et al. [2012]) but there is not a consensus in the literature as to whether larger or smaller xylem vessel diameters are ideal. Establishing, a definite answer to these issues will be important, although it should be borne in mind that different xylem diameters/configurations may be optimal for different environments. Given the predicted increasing fluctuations in rice-growing climates, genotypes exhibiting a high degree of root growth plasticity will likely increase in importance (Kano et al. [2011]). Phenotypic effects of beneficial microbes (fungi and bacteria) on plant root system include primary and lateral root length, number and positioning of lateral roots, and overall root growth (Dodd and Ruiz-Lozano [2012], Sukumar et al. [2013]). For instance, it has been shown in rice that the arbuscular mycorrhizal (AM) fungus, *Rhizophagus irregularis*, stimulates the formation of fine and large lateral roots (Gutjahr et al. [2009]). On the other hand, AM fungi form hyphal networks which operate as functional extensions of the plant root system (Smith & Read [2008]). Hence, it appears that beneficial microbes can be considered as an integral part of the root system, from which they significantly contribute to nutrient acquisition and host tolerance to occasional or prolonged abiotic and biotic stress. Our perspective is that, in addition to root architecture, future work on ideotypes should integrate functional parameters (e.g. water and ion uptake per length of root and the respective contributions of apoplast and cell-to-cell paths), plasticity, and plant/soil microbe interactions which can alter host root development and functioning.

However, as our understanding increases, the concept of root ideotype is less clearly defined. It is therefore important to exploit the power of modeling to integrate these different factors and be able to make predictions about the traits that may be beneficial. Modeling forces us to examine many parameters related to a specific process and allows us to test the sensitivity of the model to each of these. This can inform about potential rate-limiting processes that govern complex traits such as water or nutrient acquisition and guide the subsequent identification of associated QTLs. Several structural functional root models have been published (e.g. SimRoot, Lynch et al. [1997] and R-SWMS, Dunbabin et al. [2013]) as well as other soil models (Reviewed in Hill et al. [2013]). Root/soil interaction models will be important in simulating the dynamic response and efficiency of different rice ideotypes in different soil conditions. One considerable advantage of such models is that they allow us to investigate the cost-benefit relationship between various traits in a wide variety of environments. For example, producing longer roots can protect against drought but it comes at a cost of carbon that could be used elsewhere, and deep roots may be less efficient in foraging for nitrogen. Structural-functional plant models based on an economy of resources allow us to identify potential combinations of traits that can increase plant growth under a wide variety of environmental conditions, including water or nutrient poor conditions. These predictions could drive breeders or genetic engineers towards producing plants with these combinations of traits, so that the models can be tested experimentally. Again, characterization of the environment will be critical for this analysis, and having a geographically distributed network of rice root researchers with knowledge of different soils is an enormous asset. Future models need to take into account the soil structure as well as nutrient/water availability, and incorporate both root architecture and anatomy within the model. These tools will allow the search for more difficult to study root traits that are linked to adaptability, rather than fixed characters for a given environment. This effort should be community-based and where possible should use open access platforms (such as OpenAlea; Pradal et al. [2008]) to guarantee adoption and the continuity of curation of the developed models.

In summary, we think that in addition to the approach of screening for stress tolerance among diverse rice genotypes (including wild relatives), there are many benefits to be gained from developing an approach to root biology based on the integration of our knowledge of the physiology of plant response to stress to define ideotypes that can form the basis of new more specific genetic screens. For instance, our perspective is that future work on ideotypes should integrate functional parameters (i.e. water and nutrient uptake per length of root,). Future research is necessary to conceptualize new integrative root ideotypes that will enable the definition of new experimental screening approaches to contribute to rice improvement strategies. A more complex, but necessary step would be the coupling of the roots development modelling approaches and tools with the ones of the whole plant modeling in order to reason the ideotype at the whole plant level.

BREED FOR IMPROVED ROOT SYSTEMS OF FUTURE RICE VARIETIES

Several hundred QTLs for root traits (mainly structural and a few functional) have been identified and compiled for the identification of meta-QTLs (Courtois et al. [2009]; Khowaja et al. [2009]). Association mapping, which improves further the resolution in QTL position, has already been initiated for rice root traits on broad-based panels (Famoso et al. [2011]; Courtois et al. [2013]; Wissuva et al. unpublished work) and a core collection of local varieties from Vietnam (Gantet et al. unpublished work). One of these QTLs, *DRO1,* was recently cloned and its function characterized (Uga et al. [2013]). More recently, the characterization of QTLs with large effect involved in phosphorus uptake (Gamuyao et al. [2012]) and in yield under drought (Kumar et al.[2013]; Swamy et al. [2013]) revealed direct relationship with or contribution from root traits. Efforts for cloning a number of other root QTLs are ongoing at NIAS (QTLs for stele size and surface rooting), JIRCAS (QTLs for P uptake), Nagoya University (QTL for root growth in response to soil moisture fluctuations and root growth plasticity) IRRI (QTLs for yield under drought), CIRAD (QTLs for root length) and Tamil Nadu University (QTLs for root penetration ability). Thus a large number root QTLs or genes promising for improving rice performances in water and nutrient uptake are now available for breeders.

Efforts in marker assisted introgression of root QTLs from landraces into elite material go back to 2001 (Shen et al. 2001) with mixed effects on yield in target environment in some cases (Steele et al. 2007). positive effects in some other cases (Gamuyao et al. [2012]; Uga et al. [2013]; Steele et al. [2013]). An important next step for practical use of the large number of root QTLs within breeding programs would be their characterization for QTL x environment and QTL x genetic background interactions, using grain yield as the selection criterion. Indeed, root-related QTLs/genes may interact with the genetic background into which they are introgressed and with the environment to which they are targeted. Thus, it will be necessary to assess the effect of individual or combination of QTLs in the largest possible number of relevant environments and genetic backgrounds, the latest taking into account the structuration of *O. sativa* into four majors groups (*indica, japonica, aus* and *aromatic*). The systematic implementation of such an approach (all combinations of genetic backgrounds and environments) would necessitate a too large set of material to be developed. Simulation modeling could be useful to help discard non-functional trait/QTL combinations at early stages in the development of improved rice genotypes. Subsequently, the results of the yield evaluation together with existing QTL data could be consolidated to develop a breeder's chip carrying markers associated with the main QTLs and candidate genes.

We expect that as more QTLs/target genes are identified and as genetic and genomic approaches converge, a radical change in methodology to rapidly deliver new rice varieties will be necessary. Root architecture is one trait among others for breeders, and it will be necessary to concurrently manipulate or introgress in addition to genes related to root architecture, those for disease resistance, multiple abiotic stress tolerance, grain quality, etc. To deal with such complexity, an emerging solution is genomic selection using a multi-trait index (Meuwissen et al. [2001]). The methodology is being adapted to plants and shows interesting promises (Jannink et al. [2010]; Lorenz et al. [2011]). However the application of genomic selection will require making parallel progress in high throughput phenotyping to calibrate predicting models, in extraction of DNA of the right quality for a high number of plants in a short time, and decreasing even more the cost of genotyping.

To take full advantage of the potential of genomic selection, the process of recombination of multiple parents will also have to be rationalized and accelerated. The best solution to accumulate favorable alleles at multiple genes in a unique background is probably not to try to do it in one round. Since breeding is a numbers game, this would require populations that are too large. Theory demonstrated that a two or more step procedure is much more efficient although slightly slower (Kervella et al. [1993]). Recurrent selection with a male sterility gene to facilitate intercrossing will be a way to go (Grenier et al. unpublished work). As part of the favorable alleles may come from exotic material, the issue of introgression of the genes into an advanced background without disturbing the whole genetic balance has to be solved. This may mean using bridges between wild species or landraces and the elite gene pools. These approaches should help breeders in taking further advantage of the genetic diversity of root traits for improving rice for future harvests.

CONCLUSION

The Roots of Future Rice Harvests

In summary, we identified the following priority targets for rice root research during the workshop:

Have a better definition of the environmental stresses that are faced by rice crop worldwide and a better characterization of the actual (on-farm) target environments. For this and to challenge ideotype models, we must mobilize a geographically distributed network of rice root researchers with knowledge of different soils.

Create new populations and new NILs for root QTL cloning

Explore genes and alleles in wild relatives and landraces without neglecting the modern gene pools that can probably also provide interesting features that will be easier to transfer to cultivated rice.

Challenge the effect of the QTLs in several genetic backgrounds taking into account the strong species structuration, across ecosystems and then within each ecosystem. QTL expression has to be analyzed in the largest possible number of relevant situations.

Gain detailed knowledge on the roles that genes corresponding to QTLs play in root and plant physiology (i.e. mineral and water nutrition). Organize/stimulate the exchange of information between research groups aiming at identifying QTLs/having identified QTLs and groups likely to be interested by the investigation of the roles that the corresponding QTL play in root and plant physiology.

Develop tools for ortholog prediction between rice, *Arabidopsis* and crops to accelerate candidate validation and information transfer

Integrate genome editing and allele replacement tools in root QTL validation

Study how roots regulate stress responses at cell/tissue levels through cell specific *omics* approaches and cell-specific gene expression. Better understand the epigenetic regulation of roots and root development under stressful conditions Prepare for a capacity of accumulating favorable alleles at multiple genes in a unique background, notably through pyramiding and recurrent selection and establishing intermediary bridges for those residing in exotic germplasm.

Accelerate the development of dedicated high throughput phenotyping systems coupled with automated data analysis

Adopt joint reference protocols and parameters as well as priority issue questions allowing a common and shared effort of harnessing of genetic resources. Create a common phenotyping data repository for comparisons and meta-analyses.

Promote the approaches coupling complementary root ideotype based screening and selection for grain yield, each approach sustaining the other.

Design new genetic screening methods based on a better knowledge of the integrated stress responses. Integrate functional parameters (e.g. water uptake per length of root), plasticity, and plant/soil microbe interactions.

Intensify the efforts towards dynamic root/soil interaction modeling to integrate these different factors and be able to make predictions about the traits that may be beneficial under a variety of environmental conditions.

The root system being less accessible and more complex than other agronomical traits, achieving the ambitious goal of breeding the roots of the future requires a coordinated effort and joint resources. The GRiSP workshop held in Montpellier has confirmed that an expanding community, including groups previously using *Arabidopsis* as a model, is at work to tackle the characterization of the structure and function of rice roots and to integrate root traits for breeding improved varieties. This first workshop had no ambition to gather the whole community and tackle all the complex aspects of rice root research. Participants were conscious that many other groups targeting complementary aspects of root biology and ecology research in rice do exist and can join their efforts to this community. Recent advances have been accomplished in the genetic and molecular control of root development, root system phenotyping, imaging and modeling as well as in genome-wide marker assisted breeding and access to nucleotide variation in genetic resources. Altogether, this makes a unique conjunction for a community to build a promising and exciting future in rice root research. In that aim we call for a better coordination of research and breeding efforts through the creation of a strong comprehensive rice root research network as well as for appropriate support from national and international funding bodies. These efforts are both worthwhile and timely since roots are certainly one of the key traits which will help us to adjust crops to the predicted increasing climate instability and resource scarcity with the objective of feeding 9 billion people by year 2050.

Competing Interest

The authors have no competing interest.

Authors' Contributions

NA, AA, AB, BC, ACO, AD, AD, PG, AH,YI, LL, ML, DTL, BM, RM, XM, LP, AP, SR, YU, AAV, MW, JX wrote the first draft. MJB, AG, LK, HS, BS, PW added valuable comments. EG completed the manuscript writing and did the final polishing. All authors read and approved the final manuscript.

ACKNOWLEDGEMENTS

The authors thank the Global Rice Science Partnership and Agropolis Fondation (Special grant n° 1400–009 and Rhizopolis grant n° 1001–005) benefiting from a national ANR "Investissement d'Avenir" grant ANR-10-LABX-001-01) for supporting the workshop. They acknowledge the assistance of Nathalie Pivot, Cirad and Véronique Rafin, INRA in workshop organization. The root research at Cirad and University of Aberdeen is supported by the European Grant (FP7/2007-2013) under grant agreement n° 289300.27 EURoot "Enhancing resource Uptake from ROOTs under stress in cereal crops". Research at IRRI is supported by the Generation Challenge Program and the Bill and Melinda Gates Foundation. J.X. is supported by the AcRF Tier 2 grant (MOE2009-T2-1-060) from the Ministry of Education of Singapore and National Research Foundation Singapore under its Competitive Research Programme (CRP Award No. NRF2010NRF-CRP002-018). Doan Trung Luu is supported by the EU Marie Curie International

Outgoing Fellowship 'ORYZAQUA – Cell Biology of Rice Aquaporins' (PIOF-GA-2011-300150). AP acknowledges the Generation Challenge Programme funded project "Targeting drought avoidance root traits to enhance rice productivity under water limited environments". Financial support for A.G. Diedhiou was provided by the Université Cheikh Anta Diop (UCAD, VE12/13, CpVIII-Ar4) and GRISP.

*This paper is dedicated to the late memory of Pr Ping Wu who passed away in a tragic car accident on June 12th, 2014.

REFERENCES

1. Al-Shugeairy Z, Islam MS, Shrestha R, Al-Ogaidi F, Norton GJ, Price AH (2014) High throughput screening of rooting depth in rice using buried herbicide. Ann Appl Biol 165:96-107
2. Bao Y, Aggarwal P, Robbins NE II, Sturrock CJ, Thompson MC, Tan HQ, Tham C, Rodriguez PL, Vernoux T, Mooney SJ, Bennett MJ, Dinneny JR (2014) Plant roots employ a patterning mechanism to position lateral root branches towards available water. Proc Natl Acad Sci 111(25):9319-9324
3. Burton AL, Williams M, Lynch JP, Brown KM (2012) RootScan: Software for high-throughput analysis of root anatomical traits. Plant Soil 357:189-203
4. CAAS BGI IRRI, c (2014) The 3,000 rice genomes project. Giga Sci 3:7
5. Clark RT, MacCurdy RB, Jung JK, Shaff JE, McCouch SR, Aneshansley DJ, Kochian LV (2011) Three-dimensional root phenotyping with a novel imaging and software platform. Plant Physiol 156:455-465
6. Clark RT, Famoso AN, Zhao K, Shaff JE, Craft EJ, Bustamante CD, McCouch SR, Aneshansley DJ, Kochian LV (2013) High-throughput two-dimensional root system phenotyping platform facilitates genetic analysis of root growth and development. Plant Cell Environ 36:454-466
7. Coudert Y, Perin C, Courtois B, Khong NG, Gantet P (2010) Genetic control of root development in rice, the model cereal. Trends Plant Sci 15:219-226

8. Courtois B, Ahmadi N, Khowaja F, Price A, Rami JF, Frouin J, Hamelin C, Ruiz M (2009) Rice root genetic architecture: meta-analysis from a drought QTL database. Rice 2:115-128

9. Courtois B, Audebert A, Dardou A, Roques S, Ghneim-Herrera T, Droc G, Frouin J, Rouan L, Goze E, Kilian A, Ahmadi N, Dingkuhn M (2013) Genome-wide association mapping of root traits in a japonica rice panel. PLoS One 8:e78037

10. De Smet I, White PJ, Bengough AG, Dupuy L, Parizot B, Casimiro I, Heidstra R, Laskowski M, Lepetit M, Hochholdinger F, Draye X, Broadley M, Péret B, Zhang H, Hammond JP, Fukaki H, Mooney S, Lynch JP, Nacry P, Schurr U, Laplaze L, Benfey P, Beeckman T, Bennett M (2012) Analyzing lateral root development: how to move forward. Plant Cell 24:15-20

11. Dodd IC, Ruiz-Lozano JM (2012) Microbial enhancement of crop resource use efficiency. Curr Opin Biotechnol 23(2):236-242

12. Dunbabin VM, Postma J, Schnepf A, Pagès L, Javaux M, Wu L, Leitner D, Chen YL, Rengel Z, Diggle AJ (2013) Modelling root–soil interactions using three–dimensional models of root growth, architecture and function. Plant Soil 372(1–2):93-124

13. Famoso AN, Zhao K, Clark RT, Tung CW, Wright MH, Bustamante C, Kochian LV, McCouch SR (2011) Genetic architecture of aluminum tolerance in rice (Oryza sativa) determined through genome-wide association analysis and QTL mapping. PLoS Genet 7(8):e1002221

14. Fernandez R, Das P, Mirabet V, Moscardi E, Traas J, Verdeil JL, Malandain G, Godin C (2010) Imaging plant growth in 4D: robust tissue reconstruction and lineaging at cell resolution. Nat Methods 7:547-553

15. French A, Ubeda-Tomás S, Holman TJ, Bennett MJ, Pridmore T (2009) High-throughput quantification of root growth using a novel image-analysis tool. Plant Physiol 150:1784-1795

16. Gaj T, Gersbach CA (2013) ZFN, TALEN, and CRISPR/Cas-based methods for genome engineering. Trends Biotech 31(7):397-405

17. Galkovskyi T, Mileyko Y, Bucksch A, Moore B, Symonova O, Price CA, Topp CN, Iyer-Pascuzzi AS, Zurek PR, Fang S, Harer J, Benfey PN, Weitz JS (2012) GiA Roots: software for the high throughput analysis of plant root system architecture. BMC Plant Biol 12:116

18. Gamuyao R, Chin JH, Pariasca-Tanaka J, Pesaresi P, Catausan S, Dalid C, Slamet-Loedin I, Tecson-Mendoza EM, Wissuwa M, Heuer S (2012) The protein kinase Pstol1 from traditional rice confers tolerance of phosphorus deficiency. Nature 488:535-539

19. Gifford ML, Dean A, Gutierrez RA, Coruzzi GM, Birnbaum KD (2008) Cell-specific nitrogen responses mediate developmental plasticity. Proc Natl Acad Sci 105(2):803-808

20. Goodstein DM, Shu S, Howson R, Neupane R, Hayes RD, Fazo J, Mitros T, Dirks W, Hellsten U, Putnam N, Rokhsar DS (2012) Phytozome: a comparative platform for green plant genomics. Nucleic Acids Res 40:D1178-D1186

21. Gutjahr C, Casieri L, Paszkowski U (2009) Glomus intraradices induces changes in root system architecture of rice independently of common symbiosis signaling. New Phytol 182:829-837

22. Haefele SM, Nelson A, Hijmans RJ (2014) Soil quality and constraints in global rice production. Geoderma 235–236:250-259

23. Hasegawa S, Thangaraj M, O'Toole JC (1985) Root behavior: field and laboratory studies for rice and non-rice crops. In: Soil Physics and Rice. International Rice Research Institute, Los Baños, Philippines. pp 383-395

24. Henry A, Cal AJ, Batoto TC, Torres RO, Serraj R (2012) Root attributes affecting water uptake of rice (Oryza sativa) under drought. J Exp Bot 63:4751-4763

25. Hill K, Porco S, Lobet G, Zappala S, Mooney S, Draye X, Bennett MJ (2013) Root systems biology: integrative modelling across scales, from gene regulatory networks to the rhizoshere. Plant Physiol 163:1487-1503

26. Huang CF, Yamaji N, Nishimura M, Tajima S, Ma JF (2009) A rice mutant sensitive to Al toxicity is defective in the specification of root outer cell layers. Plant Cell Physiol 50:976-985

27. Inthavong T, Fukai S, Tsubo M (2011) Spatial variations in water availability, soil fertility and grain yield in rainfed lowland rice: a case study from savannakhet province, Lao PDR. Plant Prod Sci 14:184-195

28. Iyer-Pascuzzi AS, Symonova O, Mileyko Y, Hao Y, Belcher H, Harer J, Weitz JS, Benfey PN (2010) Imaging and analysis platform

for automatic phenotyping and trait ranking of plant root systems. Plant Physiol 152:1148-1157

29. Jannink JL, Lorenz AJ, Iwata H (2010) Genomic selection in plant breeding: From theory to practice. Brief Funct Genomics 9:166-177
30. Kano M, Inukai Y, Kitano H, Yamauchi (2011) Root plasticity as the key root trait for adaptation to various intensities of drought stress in rice. Plant Soil 342:189-196
31. Khalfaoui JL, Havard M (1993) Screening peanut cultivars in the field for root growth: a test by herbicide injection in the soil. Field Crops Res 32:173-179
32. Kervella J, Robert N, Fouilloux G (1993) Influence de la recombinaison sur la variabilité génétique. II. Etude par simulation. Agronomie 13:371-379
33. Khowaja FS, Norton GJ, Courtois B, Price AH (2009) Improved resolution in the position of drought-related QTLs in a single mapping population of rice by meta-analysis. BMC Genomics 10:e276
34. Kumar A, Dixit S, Henry A (2013) Marker-assisted introgression of major QTLs for grain yield under drought in rice. In: Varshney R, Tuberosa R (eds) Genomics Applications in Plant Breeding, Wiley-Blackwell Publishers, USA.
35. Lopes MS, Reynolds MP (2010) Partitioning of assimilates to deeper roots is associated with cooler canopies and increased yield under drought in wheat. Funct Plant Biol 37(2):147-156
36. Lorenz AJ, Chao S, Asoro FG, Heffner EL, Hayashi T, Iwata H, Smith KP, Sorrells ME, Jannink JL (2011) Genomic selection in plant breeding: knowledge and prospects. Adv in Agron 110:78-109
37. Lynch JP, Chimungu JG, Brown KM (2014) Root anatomical phenes associated with water acquisition from drying soil: targets for crop improvement. J Exp Bot Apr. Epub ahead of print
38. Lynch JP, Nielsen KL, Davis RD, Jablokow AG (1997) *SimRoot*: modelling and visualization of root systems. Plant Soil 188(1):139-151
39. Meijon M, Satbhai SB, Tsuchimatsu T, Busch W (2014) Genome-wide association study using cellular traits identifies a new

regulator of root development in Arabidopsis. Nat Genet 46:77-81

40. Meuwissen THE, Hayes BJ, Goddard ME (2001) Prediction of total genetic value using genome-wide dense marker maps. Genetics 157:1819-1829
41. Miao J, Guo D, Zhang J, Huang Q, Qin G, Zhang X, Wan J, Gu H, Qu LJ (2013) Targeted mutagenesis in rice using CRISPR-Cas system. Cell Res 23:1233-1236
42. Motte H, Vercauteren A, Depuydt S, Landschoot S, Geelen D, Werbrouck S, Goormachtig S, Vuylsteke M, Vereecke D (2014) Combining linkage and association mapping identifies RECEPTOR-LIKE PROTEIN KINASE1 as an essential Arabidopsis shoot regeneration gene. Proc Natl Acad Sci 111(22):8305-8310
43. Nicou R, Séguy L, Haddad G (1970) Comparaison de l'enracinement de quatre variétés de riz pluvial en présence ou absence de travail de sol. Agron Trop 25:639-695
44. Obara M, Wataru Tamura W, Ebitani T, Yano M, Sato T, Yamaya T (2010) Fine-mapping of qRL6.1, a major QTL for root length of rice seedlings grown under a wide range of NH4+ concentrations in hydroponic conditions. Theor Appl Genet 121:535-547
45. Orman-Ligeza B, Parizot B, Gantet P, Beeckman T, Bennett MJ, Draye X (2013) Post-embryonic root organogenesis in cereals: branching out from model plants. Trends Plant Sci 18:459-467
46. Ostlund G, Schmitt T, Forslund K, Kostler T, Messina DN, Roopra S, Frings O, Sonnhammer EL (2010) InParanoid 7: new algorithms and tools for eukaryotic orthology analysis. Nucleic Acids Res 38:D196-D203
47. O'Toole JC, Chang TT (1978) Drought and rice improvement in perspective. IRRI Res Pap Ser 14:27. Los Baños, Philippines
48. Pegoraro C, Farias DR, Mertz LM, dos Santos RS, da Maia LC, Rombladi CV, Costa de Oliveira A (2013) Ethylene response factors gene regulation and expression profiles under different stresses in rice. Theor Appl Genet 25:261-274
49. Pennisi E (2010) Sowing the seeds for the ideal crop. Science 327(5967):802-803

50. Postma JA, Dathe A, Lynch JP (2014) The optimal lateral root branching density for maize depends on nitrogen and phosphorus availability. Plant Physiol. [Epub ahead of print]
51. Pound MP, French AP, Atkinson JA, Wells DM, Bennett MJ, Pridmore T (2013) RootNav: navigating images of complex root architectures. Plant Physiol 162:1802-1814
52. Pradal C, Dufour-Kowalski S, Boudon F, Fournier C, Godin C (2008) OpenAlea: a visual programming and component-based software platform for plant modeling. Funct Plant Biol 35:751-760
53. Ranathunge K, Steudle E, Lafitte R (2003) Control of water uptake by rice (Oryza sativa L.): role of the outer part of the root. Planta 217:193-205
54. Ranathunge K, Kotula L, Steudle E, Lafitte R (2004) Water permeability and reflection coefficient of the outer part of young rice roots are differently affected by closure of water channels (aquaporins) or blockage of apoplastic pores. J Exp Bot 55:433-447
55. Rebouillat J, Diévart A, Verdeil JL, Escoute J, Giese G, Breitler JC, Gantet P, Espeout S, Guiderdoni E, Périn C (2009) Molecular genetics of rice root development. Rice 2(1):15-34
56. Rouard M, Guignon V, Aluome C, Laporte MA, Droc G, Walde C, Zmasek CM, Perin C, Conte MG (2011) Green) PhylDB v2.0: comparative and functional genomics in plants. Nucleic Acids Res 39:D1095-D1102
57. Schreiber F, Sonnhammer EL (2013) Hieranoid: hierarchical orthology inference. J Mol Biol 425:2072-2081
58. Shen L, Courtois B, McNally KL, Robin S, Li Z (2001) Evaluation of near-isogenic lines of rice introgressed with QTLs for root depth through marker-aided selection. Theor Appl Genet 103:75-83
59. Smith SE, Read DJ (2008) Mycorrhizal Symbiosis. Academic Press, New York.
60. Steele KA, Price AH, Witcombe JR, Shrestha R, Singh BN, Gibbons JM, Virk DS (2013) QTLs associated with root traits increase yield in upland rice when transferred through marker-assisted selection. Theor Appl Genet 126:101-108

61. Steele KA, Virk DS, Kumar R, Prasad S, Witcombe JR (2007) Field evaluation of upland rice line selected for QTLs controlling root traits. Field Crop Res 101:180-186

62. Sukumar P, Legué V, Vayssières A, Martin F, Tuskan GA, Kalluri UC (2013) Involvement of auxin pathways in modulating root architecture during beneficial plant–microorganism interactions. Plant Cell Environ 36:909-919

63. Swamy BPM, Ahmed HU, Henry A, Mauleon R, Dixit S, Vikram P, Ram T, Verulkar SB, Perraju P, Mandal NP, Variar M, Robin S, Chandrababu R, Singh ON, Dwivedi JL, Das SP, Mishra KK, Yadaw RB, Aditya TL, Karmakar B, Satoh K, Moumeni A, Kikuchi S, Leung H, Kumar A (2013) Genetic, physiological, and gene expression analyses reveal that multiple QTL enhance yield of rice mega-variety IR64 under drought. PLoS One 8:e62795. doi:10.1371/journal.pone.0062795

64. Topp CN, Iyer-Pascuzzi AS, Anderson JT, Lee CR, Zurek PR, Symonova O, Zheng Y, Bucksch A, Mileyko Y, Galkovskyi T, Moore BT, Harer J, Edelsbrunner H, Mitchell-Olds T, Weitz JS, Benfey PN (2013) 3D phenotyping and quantitative trait locus mapping identify core regions of the rice genome controlling root architecture. Proc Natl Acad Sci 110:E1695-E1704

65. Trachsel S, Kaeppler SM, Brown KM, Lynch JP (2011) Shovelomics: high throughput phenotyping of maize (*Zea mays* L.) root architecture in the field. Plant Soil 341:75-87

66. Tracy SR, Roberts JA, Black CR, McNeill A, Davidson R, Mooney SJ (2010) The X-factor: visualizing undisturbed root architecture in soils using X-ray computed tomography. J Exp Bot 61:311-313

67. Uga Y, Sugimoto K, Ogawa S, Rane J, Ishitani M, Hara N, Kitomi Y, Inukai Y, Ono K, Kanno N, Inoue H, Takehisa H, Motoyama R, Nagamura Y, Wu J, Matsumoto T, Takai T, Okuno K, Yano M (2013) Control of root system architecture by *DEEPER ROOTING 1* increases rice yield under drought conditions. Nat Genet 45:1097-1102

68. Wang LK, Chu HW, Li ZY, Wang J, Li JT, Qiao Y, Fu YR, Mou TM, Chen CL, Xu J (2014) Origin and development of the root cap in rice. Plant Physiol 114.240929. Epub ahead of print

69. Wei FJ, Droc G, Guiderdoni E, Hsing YI (2013) International consortium of rice mutagenesis: resources and beyond. Rice (N Y) 6:39

70. Woll K, Borsuk LA, Stransky H, Nettleton D, Schnable PS, Hochholdinger F (2005) Isolation, characterization, and pericycle-specific transcriptome analyses of the novel maize lateral and seminal root initiation mutant rum1. Plant Physiol 139:1255-1267

71. Yambao EB, Ingram KT, Real JG (1992) Root xylem influence on the water relations and drought resistance of rice. J Exp Bot 43:925-932

Chapter 3

Global Expression Profiling of Low Temperature Induced Genes in the Chilling Tolerant Japonica Rice Jumli Marshi

[1]Aakash Chawade,[2] Angelica Lindlöf,[3] Björn Olsson and [4]Olof Olsson.

[1]CropTailor AB, Lund, Sweden

[2]CropTailor AB, Lund, Sweden, Systems Biology Research Centre, School of Life Sciences, University of Skövde, Skövde, Sweden

[3]Systems Biology Research Centre, School of Life Sciences, University of Skövde, Skövde, Sweden

[4]CropTailor AB, Lund, Sweden, Department of Pure and Applied Biochemistry, Lund University, Lund, Sweden

ABSTRACT

Low temperature is a key factor that limits growth and productivity of many important agronomical crops worldwide. Rice *(Oryza sativa* L.) is negatively affected already at temperatures below +10°C and is therefore denoted as chilling sensitive. However, chilling tolerant rice cultivars exist and can be commercially cultivated at altitudes up to 3,050 meters with temperatures reaching as low as +4°C. In this work, the global transcriptional response to cold stress (+4°C) was studied in the Nepalese highland variety Jumli Marshi (spp. *japonica*) and 4,636 genes were identified as significantly differentially expressed within 24 hours of cold stress. Comparison with previously published microarray data from one chilling tolerant and two sensitive rice cultivars identified 182 genes differentially expressed (DE) upon cold stress in all four rice cultivars and 511 genes DE only in the chilling tolerant rice. Promoter analysis of the 182 genes suggests a complex cross-talk between ABRE and CBF regulons. Promoter analysis of the 511 genes identified over-represented ABRE motifs but not DRE motifs, suggesting a role for ABA signaling in cold tolerance. Moreover, 2,101 genes were DE in Jumli Marshi alone. By chromosomal localization analysis, 473 of these cold responsive genes were located within 13 different QTLs previously identified as cold associated.

INTRODUCTION

The inherent ability of plants to endure low temperatures affects both their geographical distribution and overall productivity. Many cereal crops from temperate regions, e.g., wheat, barley, rye and oat, have evolved efficient protection mechanisms to tolerate freezing [1]. This tolerance is acquired through a process known as cold acclimation, which occurs at low, but above-zero temperatures [2]. On the other hand, the highly important staple food, rice, like many other tropical plants, is chilling sensitive and does not survive freezing temperatures even after cold acclimation. However, chilling susceptibility varies substantially among cultivars of the same species [3], indicating that even chilling sensitive crops, at least to a certain degree, can acclimate to cooler temperatures. For example, *japonica* rice cultivars show a greater tolerance to chilling than *indica* cultivars, although variations within *japonica* exist [4]. Rice is mainly grown in warm climates (>25°C), but exposure to low temperature (LT) is common for rice cultivated in temperate zones or at high elevations in several regions of Europe, South Asia and Southeast Asia. It is, therefore, of outstanding scientific and economic interest to understand chilling tolerance, identify key regulatory components of chilling acclimation, resolve mechanistic differences between tolerant and sensitive rice and apply this knowledge in the development of new, more tolerant rice cultivars.

It is known that cold acclimation leads to physiological and metabolic changes in cell and tissue structures as a result of an extensive reprogramming in gene expression [2], [5], [6], [7]. A large number of genes that are differentially expressed during cold acclimation have been identified and characterized in important cold hardy cash crops like wheat (*Triticum aestivum*)[8], barley (*Hordeum vulgare*) [9], [10] and oat (*Avena sativa*) [7], [11]. However, the plant species most extensively studied during cold stress and acclimation is a non-crop plant,*Arabidopsis thaliana,* where global transcriptional profiling experiments have identified several cold responsive genes [12], [13], [14], [15], [16], [17]. The complexity of genetic re-programming upon cold stress has also been demonstrated by various bioinformatics approaches [15], [18], [19], [20], [21].

Previously, Rabbani *et al.* [22] used a rice cDNA microarray of 1,718 ESTs and identified 36 cold responsive genes in two-week old seedlings of chilling tolerant rice Nipponbare (spp.*japonica*) exposed to +4°C for 24 hours. Cheng *et al.* [23] used a rice cDNA microarray of 5,855 unique ESTs and identified 121 cold responsive genes in 10 days old seedlings of the chilling tolerant rice CT6748-8-CA-17 (spp. *japonica*) treated at +10°C for up to 24 hours. In another study, Oda *et al.* [24] used the 44K Agilent oligonucleotide microarray to compare two *japonica*cultivars Sasanishiki and Hitomebore exposed to low temperature stress (19°C) at the reproductive stage. Microarray analysis of anthers from the two cultivars led to the identification of 356 differentially expressed genes in either or both cultivars. Yun *et al.* [25] analyzed genes induced by chilling stress (+10°C) in Nipponbare using microarrays representing 40,000 genes and identified 8,668 differentially expressed genes. Mittal et al. [26] performed microarray analysis of the *indica* rice Pusa Basmati that was cold stressed at +5°C and identified 924 differentially expressed genes. Zhang et al. [27] performed comparative microarray analysis of a chilling tolerant rice cultivar (LTH; *japonica*) and a chilling sensitive rice cultivar (IR29; *indica*) and showed that although the early response to low temperatures was similar in the two cultivars, the genes that were expressed at the later time points belonged to substantially different functional categories. These studies clearly show that chilling tolerance varies among rice cultivars and that many rice genes respond to low temperature stress.

In order to gain new insights into cold stress response in rice, the main aim of this study was to conduct a global cold (+4°C) responsive gene expression profiling of the Nepalese highland rice cultivar Jumli Marshi (JM) (spp. *japonica*). This rice is grown in the district Jumla located at an altitude of up to 3,050 m in Nepal. The average maximum and minimum temperature in the region is +21°C and +4°C, respectively [28]. JM has been the preferred variety grown in this district for several decades, thus making Jumla the highest and coldest region in the world for commercial cultivation of rice. The fact that JM can be grown at this altitude while maintaining productivity corresponding to 60% of the average Nepal rice productivity per hectare [28] implies that JM is indeed chilling tolerant.

Obviously, JM has developed ways to protect itself from cold stress. Identifying genes involved in the underlying molecular mechanisms could reveal new insights into how cold tolerance is achieved. This will ultimately enable development of new cultivars with improved cold tolerance. In this work, we performed transcriptome analysis of JM under cold stress and identified 4,636 differentially expressed genes. To our understanding, this is the first report on the global analysis of cold responsive genes in Nepalese highland rice during early cold stress (<24 h, +4°C).

MATERIALS AND METHODS

Plant Material and Growth Conditions

***Regular growth condition*:** Seeds of two rice cultivars, *Oryza sativa*, ssp. *japonica*, cv. Jumli Marshi (JM) and ssp. *indica*, cv. IR64 (IR64), were first soaked in water for 16 hours at room temperature and thereafter grown on standard soil in 14 hours photoperiod, with a day/night air temperature of 25°C/20°C and 250 μmol m^2 s^{-1} light.

Cold Condition

At mid-day, three weeks old plants were transferred to growth chambers (Percival) in the same photoperiod, but with an air temperature of 4°C and a light intensity of 100 μmol m^2 s^{-1}. Pooled leaf tissue from five individual Jumli Marshi plants were harvested at 0, 0.5, 2, 4, 8, and 24 hours, frozen in liquid nitrogen and stored at −80°C until further analysis.

Chlorophyll Fluorescence Measurements

Chlorophyll fluorescence was measured with the portable chlorophyll fluorometer PAM 2000 (Heinz Walz GmbH, Germany) and the photosystem II efficiencies $F_v/F_m = (F_m - F_0)/F_m$ were estimated as per the manufacturer's instructions. Plants were dark acclimated for one hour before taking measurements from up to 20 individual seedlings.

Quantitative Real-time RTPCR Measurements

TotalRNA was extracted with RNeasy plant mini kit (Qiagen, Cat. No. 74904) as per the manufacturer's instructions. DNAse digestion was performed on-column as per the instructions using RNase-Free DNase Set (Qiagen, Cat. No. 79254). The primer sets used in the study were DREB1aF 5'GGACCTGTACTACGCGAGCTT3', DREB1aR 5'GGG AAA ATT GTA CAG TTG ATT GA3', DREB1bF 5'AGC TCG CCG GCT CCG ACA3', DREB1bR 5'GGG AGA ATT CTG GCA CAT TCC3', DREB1cF 5'GAG TTG GAG CTA GCA GTT TTG AG3', DREB1cR 5'TAG CTG TAT AGG AGG AGC AAA GC3', OsActin1F 5'ATC CTT GTA TGC TAG CGG TCG A3', OsActin1R 5'ATC CAA CCG GAG GAT AGC ATG3' [29], [30]. Quantitative real-time PCR was performed on biological triplicates with the default protocol in iScript One-Step RT-PCR kit with SYBR Green kit (BioRad, Cat. No. 170-8893) using BioRad C1000Thermal Cycler. Relative expression of the genes was calculated with the Pfaffl method [31].

Data Preparation and Analysis

Total RNA was extracted from JM leaf tissue with TRIZOL reagents (Invitrogen) according to the manufacturer's protocol and purified by RNeasy MinElute Cleanup Kit (Qiagen). The RNA quality and concentration was measured using Agilent 2100 BioAnalyzer and Nanodrop ND-1000. Biotinylated target cRNA was prepared from 4 μg of total RNA following the manufacturer's specifications (Affymetrix). The samples were then hybridized to Affymetrix GeneChip® Rice Genome Arrays, which contain probes to query ~51K transcripts representing both *japonica* and *indica* cultivars. The chips were thereafter washed and stained in a GeneChip® Fluidics Station 450. Scanning was carried out with GeneChip® Scanner 3000 and image analysis was performed using GeneChip® Operating Software. Two biological replicates were analysed per time point. The CEL files were submitted to ArrayExpress with the accession number E-MEXP-3718.

Data was processed using Bioconductor [32] in R v2.14. Raw CEL files were background corrected with the GCRMA method and quantile normalized using the Bioconductor package affyPLM v1.30 [33]. Probe-set present/absent calls were calculated with the mas5calls method in the AffyBatch package [34]. Probe sets that were marked present in at least one of the samples were considered for further analysis. Probe sets with IQR greater than the median IQR of all probe sets were selected with the genefilter v1.36 package. To identify differentially expressed probe sets, a linear contrast matrix was built between the control and the cold-treated samples using the Limma v3.10 package [35]. Differentially expressed genes (DEGs) were identified with the empirical Bayes method in the Limma package using default parameter settings. A Benjamini-Hochberg corrected p-value <0.05 was set as the significance threshold. Thereafter, only probe sets with a unique RAP OS ID were retained for further analysis.

MapMan annotations for Oryza sativa (spp. *japonica*) genes (v1.0) were downloaded (www.mapman.gabipd.org), and annotations for the DEGs were extracted using a custom Perl script. Hierarchical clustering was performed with Cluster 3.0, using the average linkage method and Pearson correlation as similarity measure. Dendrograms were generated with Java TreeView 1.1.6. Clustering was also performed with Short Time-series Expression Miner (STEM, version 1.3.8) [36] using default parameters.

RESULTS

JM has Higher Cold Tolerance than IR64

To confirm the higher cold tolerance in JM compared to lowland rice, three weeks old seedlings grown at regular conditions were transferred to cold conditions and exposed to +4°C for three days (see Materials and Methods). Plants were then allowed to recover for two weeks in regular growth conditions, and their viability was visually estimated. The results confirmed that JM is chilling tolerant while IR64 is not since all IR64 plants wilted and died while the JM plants recovered from the cold stress (Figure 1a, 1b).

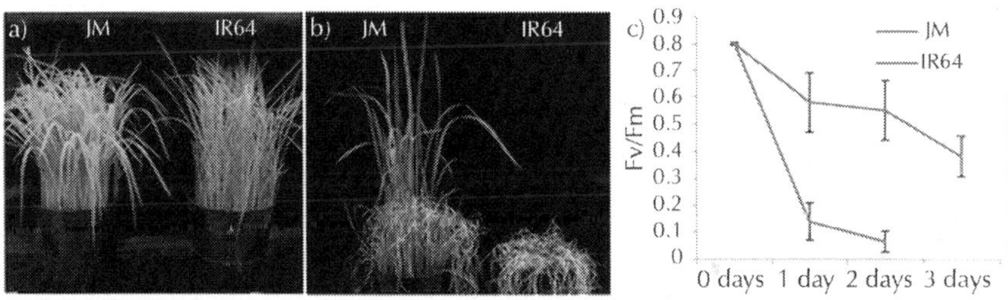

Figure 1: Cold stress survival in Jumli Marshi (JM) and IR64. Jumli Marshi (JM) and IR64 seedlings were grown for three weeks under regular growth conditions (see Materials and Methods) and then moved to +4°C (cold conditions).

After three days in cold conditions, plants were moved back to regular growth conditions and allowed to recover for two weeks. (a) Plants just before cold exposure; (b) Cold treated plants after recovery for two weeks; JM, left; IR64, right; (c) Chlorophyll fluorescence in JM and IR64 undergoing cold stress.

doi:10.1371/journal.pone.0081729.g001.

To quantify the stress levels in the two cultivars, chlorophyll fluorescence signals were measured. The ratio between the variable (F_v) and maximum (F_m) fluorescence signals reflects the efficiency of photosystem II. Plants were first dark-acclimated for 1 hour, and the fluorescence was measured (PAM 2000 fluorometer; Heinz Walz GmbH, Germany) just before transferring the plants to +4°C (0 days), and then after 1, 2 and 3 days in the cold. Readings from 50 individual seedlings were collected, and the experiment was repeated twice. Prior to cold exposure, F_v/F_m ratios were similar (~0.8) in both JM and IR64 plants, indicating that all plants were physiologically healthy. After one day in cold, the F_v/F_m ratio in IR64 fell to 0.14, while, in JM, the decrease was relatively small (0.64). The differences between the averages were significant at $p<0.0001$ (Student's t-test). After two days in cold, the F_v/F_m ratio reached near zero in IR64, while, in JM, it reached 0.55 (Figure 1c). These results show that JM tolerates cold stress significantly better than IR64.

Identification of Differentially Expressed Genes under Cold Stress in JM

Since JM is chilling tolerant, it was chosen for global gene expression profiling under cold stress. Three weeks old plants grown under regular conditions were moved to cold conditions (+4°C) at mid-day and pooled leaf tissues from five individual plants were harvested at six different time points: 0, 0.5, 2, 4, 8 and 24 hours.

Microarray data analysis revealed that 4,636 genes were significantly differentially expressed (Table S1). The complex transcriptional response pattern to early cold stress in JM can be visualized by hierarchical clustering (Figure 2). Two major clusters were formed, one with 1,490 up-regulated genes, and the other with 3,146 down-regulated genes. There were 183 genes (148 up, 35 down) DE by at least two folds ($\log_2$) within two hours of cold exposure. Most common annotations in this set were genes with unknown function (83 genes), regulation of transcription (36 genes), protein degradation or post-translational modification (16 genes) and signaling (8 genes) (Figure S1). Out of the 36 genes with transcription factor (TF) activity, the most common TF families were AP2/EREBP (12 genes), WRKY (5 genes) and bHLH (4 genes). The most highly induced genes within 2 hours were *OsDREB1B* (10 $\log_2$ folds) and*OsDREB1A* (8.9 $\log_2$ folds). Quantitative real-time RT-PCR measurements of six genes (Figure 3 and Figure S2) showed good correlation with the microarray data.

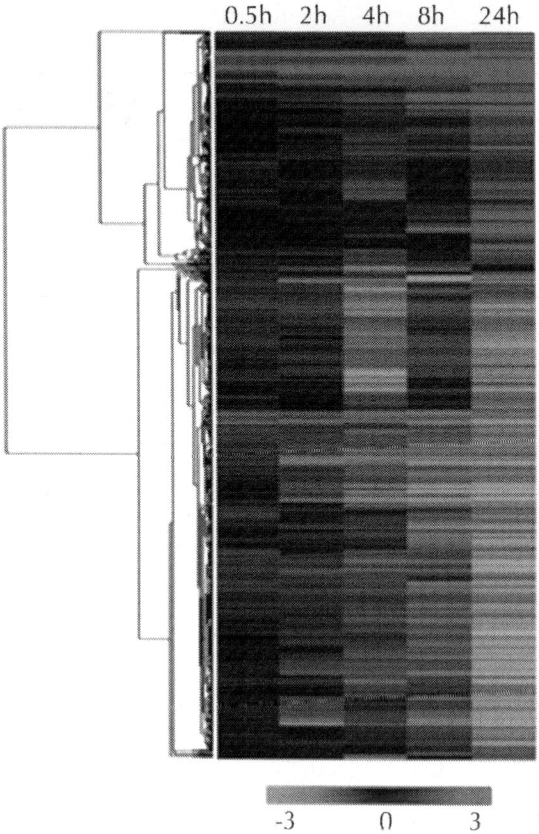

Figure 2: Hierarchical clustering of JM genes significantly differentially expressed in at least one time point under cold stress.

Clustering was performed with Cluster 3.0, using the average linkage method and Pearson correlation as similarity measure. The tree was generated with Java TreeView 1.1.6. Red: up-regulated genes; green: down-regulated genes. Log_2 transformed fold change expression levels are indicated by the color gradient. The extreme values in the color gradient are −3 to +3 in log_2 scale.

doi:10.1371/journal.pone.0081729.g002.

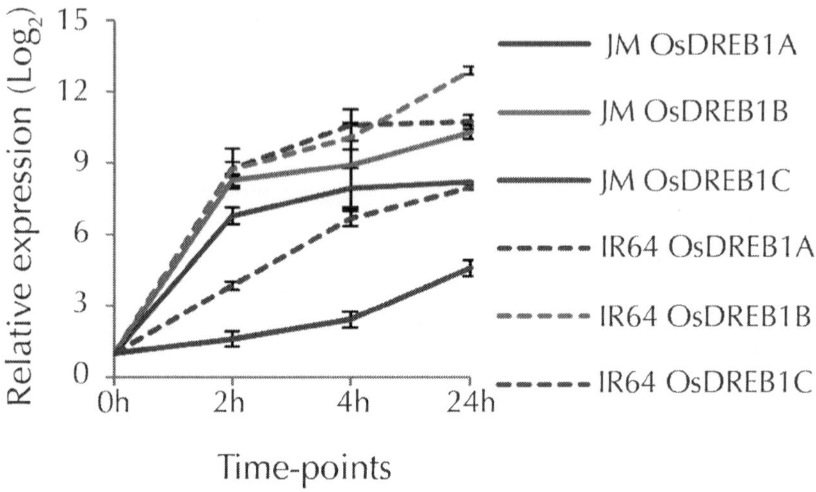

Figure 3: Monitoring Os*DREB1* mRNA levels in JM and IR64 by quantitative real-time RT-PCR.

Relative expression levels of Os*DREB1a*, OS*DREB1b* and Os*DREB1c* at 0 h, 2 h, 4 h and 24 h measured by quantitative real-time RT-PCR measured in biological triplicates. Solid line represents JM and dashed line represents IR64.

doi:10.1371/journal.pone.0081729.g003.

DREB1 Genes are Similarly Expressed in JM and IR64

*DREB1*s encode transcription factors that belong to the AP2 family of DNA binding proteins. Several experiments in many different plant species have shown that *DREB1A, DREB1B* and*DREB1C* are induced within 30 minutes of exposure of plants to cold. For example, over-expression of *DREB1*s in *Arabidopsis* resulted in increased freezing tolerance of non-acclimated transgenic plants compared to the non-transformed control plants [37]. Thus, the mRNA expression levels of the three *DREB1* genes were measured by quantitative RT-PCR in JM and IR64. The results revealed that all three genes were highly induced (>5 folds) within 2 hours in both JM and IR64, although, the expression levels were higher in IR64 than in JM (Figure 3). Previously Zhang et al [38] identified 22 genes as part of the OsDREB1c regulon. In JM, 18 of these 22 genes were DE upon cold stress indicating the critical role of OsDREB1c regulon under cold tolerance in JM (Figure S3).

Comparative Analysis of Microarray Data from Four Rice Cultivars Reveals Genes that are Differentially Expressed only in Chilling Tolerant Rice

Comparative analysis with previously published rice microarray data was done to identify genes that are differentially expressed (DE) only in chilling tolerant rice. Zhang *et al.* [27] performed microarray data analyses of the chilling tolerant rice cultivar LTH (spp. *japonica*) and the chilling sensitive rice cultivar IR29 (spp. *indica*). Mittal *et al.* [26] performed microarray data analysis of the rice cultivar Pusa Basmati (PB1, ssp. *indica*). PB1 is mostly cultivated in the North-Western part of India [39] in the warm season May-November and thus is considered chilling sensitive.

In this work, comparative analysis was done between JM and the three previously published datasets and the genes that were DE by at least 2 folds in at least one of the four rice cultivars were studied further (Figure 4). The analysis identified 182 genes that were DE by at least two folds in all four rice cultivars. These genes were termed Common Cold Induced (CCI) (Figure 4,Table 1). The CCI gene list includes several well-known cold induced genes induced by 2 folds within 2 hours after induction including OsDREB1C, OsDREB1G, OsWRKY71, OsNAC3 and also genes induced later than 2 hours including OsWRKY1 and OsNAC4. The comparative analysis also identified 511 genes that were only expressed in the two chilling tolerant cultivars JM and LTH. These 511 genes were termed as Cold Induced in Tolerant cultivars (CIT). Expression levels of three genes from the CIT group were validated by quantitative RT-PCR (Figure S2). As the CIT genes are DE only in the chilling tolerant cultivars, they may play a significant role in chilling tolerance in rice. A third set with 2,102 genes was only DE in JM. These were denoted as JM Only (JMO). As JM is cultivated at altitudes of up to 3,050 m, further understanding of the JMO genes may help to identify novel mechanisms for survival of plants in these environmental conditions. Therefore, the JMO genes were further analyzed in this study. A complete list of genes in different groups is in Table S2.

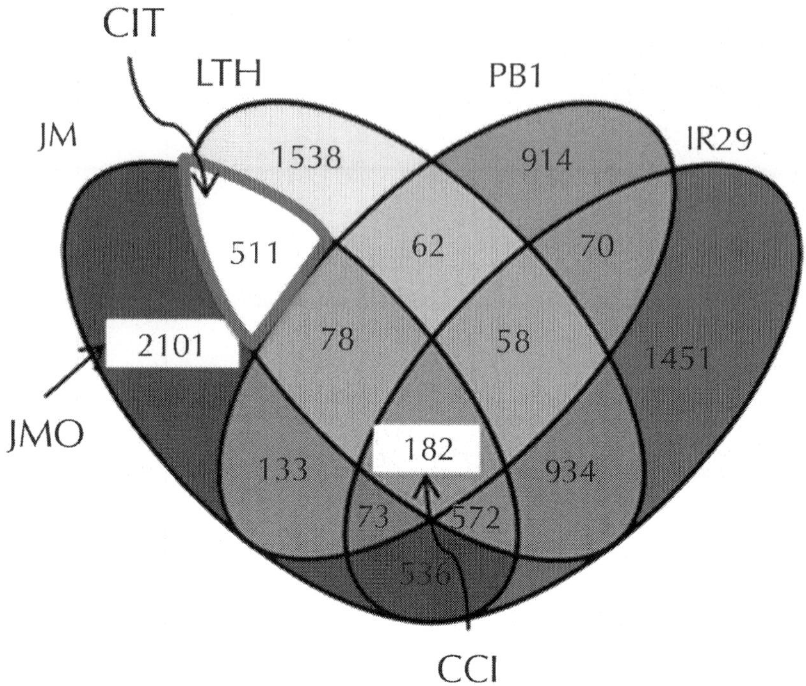

Figure 4: Number of DE genes in four rice cultivars.

Comparison of DE genes in four rice cultivars upon cold stress. CCI (Common Cold Induced), CIT (Cold Induced in Tolerant cultivars), JMO (Jumli Marshi Only). Venn diagram created using VENNY [71].

doi:10.1371/journal.pone.0081729.g004

Table 1: CCI (Common Cold Induced) genes differentially expressed in all four rice cultivars

doi:10.1371/journal.pone.0081729.t001

Os_RAPDB_ID	0.5 h	2 h	4 h	8 h	24 h	Symbol	Bin	MapMan description
Os02g0677300	5.4	5.7	6.3	7.6	7.6	OsDREB1G	27.3.3	RNA.regulation of transcription.AP2/EREBP
Os01g0186900	5.1	6.9	7.3	7.9	9		35.2	not assigned.unknown
Os10g0391400	4.9	4.7	5.7	7	6.5		35.2	not assigned.unknown
Os08g0482600	4.8	3.9	2.9	6.6	5.4		26.19	misc.plastocyanin-like
Os12g0150200	4	3.7	5.1	7.3	7.2		26.1	misc.cytochrome P450
Os10g0562900	3.9	5.1	5.6	6	6.6		27.3.3	RNA.regulation of transcription.AP2/EREBP
Os01g0955100	3.9	4.6	4.5	5.2	6.2		30.3	signalling.calcium
Os02g0629000	3.8	4	3.9	5.4	7.5		35.2	not assigned.unknown
Os02g0134200	3.6	5.4	6.2	6.9	8.4		35.2	not assigned.unknown
Os02g0703600	3.4	5.5	6.3	7.4	8.5	OsABA8OX1	26.1	misc.cytochrome P450
Os02g0181300	3.2	3.5	4	5.2	5.8	OsWRKY71	27.3.32	RNA.regulation of transcription.WRKY
Os04g0301500	3.2	2.1	2.9	5.4	5.4	27.3.6		RNA.regulation of transcription.bHLH,Basic Helix-Loop-Helix
family								
Os03g0741100	3.1	2	2.4	4.7	4.6	OsbHLH148	27.3.6	RNA.regulation of transcription.bHLH,Basic Helix-Loop-Helix

Os01g0246700	1.9	0.8	1.1	3	4	OsWRKY1	27.3.32	RNA.regulation of transcription.WRKY
Os05g0545400	1.9	4.6	4.4	6.5	7.8		29.4	protein.postranslational modification
Os03g0188500	1.9	2.8	3.8	5.2	6.8		35.2	not assigned.unknown
Os05g0380900	1.8	3.6	3.8	4.1	4.9		30.3	signalling.calcium
Os01g0905200	1.8	2.5	3.2	4.5	6.1		31.4	cell.vesicle transport
Os04g0497000	1.8	2.9	3.1	4.7	5.4		26.7	misc.oxidases - copper, flavone etc.
Os03g0180800	1.8	3.1	4.5	5.1	5.4	OsJAZ3	35.2	not assigned.unknown
Os06g0662200	1.7	0.7	1.3	1	3.1		27.3.35	RNA.regulation of transcription.bZIP transcription factor
family								
Os04g0372700	1.7	3.5	3.8	4.7	5.9		26.7	misc.oxidases - copper, flavone etc.

Identification of over-represented cis-elements in CCI and CIT gene sets.

Clustering was performed with STEM [36] to identify the CCI and CIT genes that were significantly co-expressed. The clustering of CCI genes resulted in two statistically significant clusters with 127 and 14 genes each, whereas the CIT genes were grouped into six clusters (Figure 5). Promoter analyses of genes in the eight clusters by the Osiris tool [40] lead to the identification of over-represented cis-elements/motifs previously known to be involved in cold regulation (Table 2). For example, CCI cluster 1 contains the well-known cold regulated motifs ABRE and DRE/CRT, and eleven other previously known motifs. This suggests a cross-talk between various transcription factor families during cold stress. Among the six CIT clusters, known over-represented motifs were only identified in cluster 3. In the remaining five clusters, no previously known motifs could be identified, suggesting that yet unknown regulons are involved in regulating many of the CIT genes.

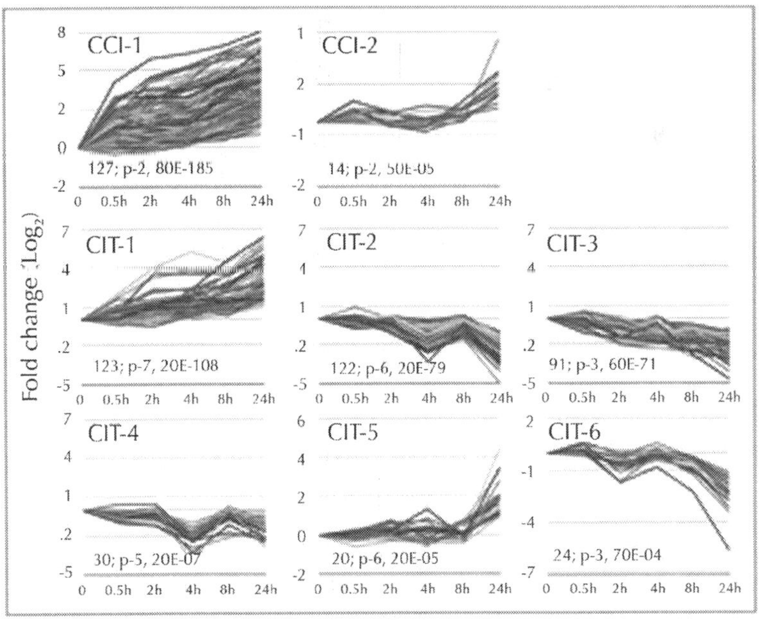

Figure 5: STEM clustering of CCI and CIT gene sets.

STEM clustering identified two significant clusters in the CCI gene set and six significant clusters in the CIT gene set.

doi:10.1371/journal.pone.0081729.g005

Table 2: Previously known significantly over-represented ($p<0.01$) cis-elements/motifs in CCI (Common Cold Induced) and CIT (Cold Induced in Tolerant cultivars) clusters.

doi:10.1371/journal.pone.0081729.t002

Cis-element	**Consensus** sequence	#Pro	p	Description
CCI Cluster 1				
ABREOSRAB21	ACGTSSSC	52	10^{-5}	ABA responsive element (ABRE) of wheat Em and rice (Os) rab21 genes
mourn	TACGTGTC	14	10^{4}	Similar to ABRE element
DRECRTCOREAT	RCCGAC	83	10^{-3}	Core motif of DRFJCRT cis-acting element
ACGTABREM011-FA2OSEM	ACGTGKC	58	10^{-3}	ACGT-core of motif A in ABRE of the rice gene
CGACGOSAMY3	CGACG	105	10^{3}	CGACG element found in the GC-rich regions of the rice
G-box-like	CACGTG	56	10^{-3}	Found in rice sucrose phosphate synthase gene promoter
ACGTOSGLUB1	GTACGTG	34	10^{3}	Required for endosperm-specific expression
GCrichrepeatll	CGCCGCGC	30	10^{-3}	Found in RAB21 promoter
MYB1AT	WAACCA	117	10^{3}	Found in promoter of chymotrypsin inhibitor-like 1 gene

ABREmotif	TGACGT	63	10^{3}	Found in promoter of 6-phosphogluconate Dehydrogenase;
SITEIOSPCNA	CCAGGTGG	12	10^{3}	Site] of rice (Os.) PCNA (proliferating cell nuclear antigen) gene
CCI Cluster 2				
E2F 1 OSPOIA	GCGGGAAA	4	10^{3} Involved in transcriptional activation in actively dividing cells and tissue	
ACGTABREM011FA2OSEM ACGTGKC		10	10^{-3}	
CIT Cluster 3				
ABRE3OSRA816	GTACGTGGCGC	2	10^{3}	ABA-responsive element of rice
MYCATERD1	CATGTG	69	10^{-2}	Found in promoter of chymotrypsin inhibitor-like 1 gene
MYCATRD22	CACATG	69	10^{3}	Found in promoter of chymotrypsin **inhibitor-like 1 gene**
WUSATAg	TTAATGG	28	10^{3}	Target sequence of WUS in the intron of AGAMOUS gene in Arabidopsis

To identify new putative motifs, gene promoters from CIT clusters 1, 2, 4, 5 and 6 were analyzed using the Element algorithm [41]. The 1 kb upstream region was analyzed for each promoter, and the top five hits were retained (Table 3). This analysis showed that while a subset of the over-represented words were entirely or partially similar to previously known cis-elements, the remaining words did not belong to any previously known cis-elements and thus are newly identified putative cis-elements (Table S3).

Table 3: Over-represented words in the 1 kb upstream promoter regions were identified using the Element algorithm [41].

doi:10.1371/journal.pone.0081729.t003

Group	Words	Known cis elements
CIT Cluster 1		
1	6	ICAT, SORUR1AT. ABRELATEER01, ACGTATE101, BOX11OCCHS
2	22	ABRELATERD1, ACGTATERD1, CACGTGAIOTIF, 1110205, T/GBOXATRIN2
3	26	POLLEN1LELAT52, GTIGAISCA/44. CGCCORE231. TATABOX5
4	13	CURECORKR
5	9	None
CIT Cluster 2		
1	25	GT1G/A5C-0,164, DOKOREZIA
2	16	WBOXATNPR1, WRIC17105
3	11	None
4	9	None
5	13	!BOX
CIT Cluster 4		
1	2	None
2	6	SCRUP2AT, ABIOS, HYSAT
3	6	None
4	4	None
CIT Cluster 5		
1	2	None
2	9	None
3	4	None
4	4	None
5	1	None
CIT Cluster 6		
1	16	None
2	IS	ACGTATE101
3	7	DOFCORE7.31,TAAAGST/511, ACGTATFJ1D1
4	11	GT1G/ASCJOA4
5	8	None

Comparative analysis of functional annotations

MapMan annotations common in the CCI and CIT gene sets were identified. Examples of such annotations are regulation of transcription, post-translational modification, protein degradation, development and calcium signaling. Interestingly, the CIT gene set also included additional annotations such as signaling in receptor kinases, protein synthesis and RNA binding (Figure 6). MapMan annotations were also obtained for all 4,636 DEGs (Table S4). The top three annotations were protein (645), RNA (574) and miscellaneous (227) (Figure 7). Analysis of annotations that were highly abundant in JM and LTH compared to chilling sensitive PB1 and IR29 showed that amino acid metabolism, RNA transcription, transport of sugars, photosynthesis light reaction and secondary metabolisms were the top five over-represented annotations in the cold tolerant cultivars (Figure 8).

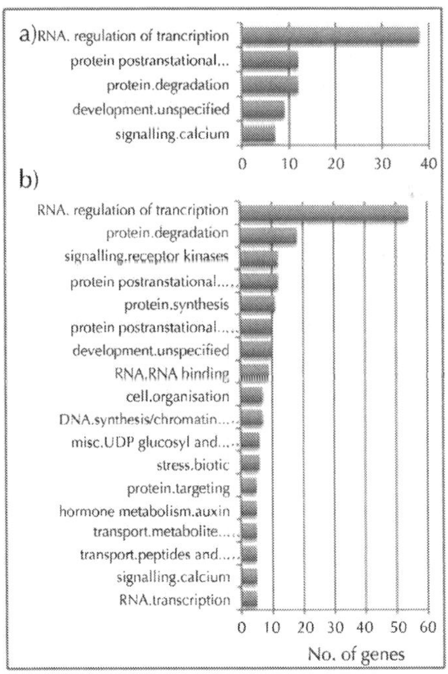

Figure 6: Functional annotations in CCI and CIT gene sets MapMan annotations in CCI and CIT gene sets. Categories with at least two genes are shown. a) CCI gene set; b) CIT gene set. doi:10.1371/journal.pone.0081729.g006.

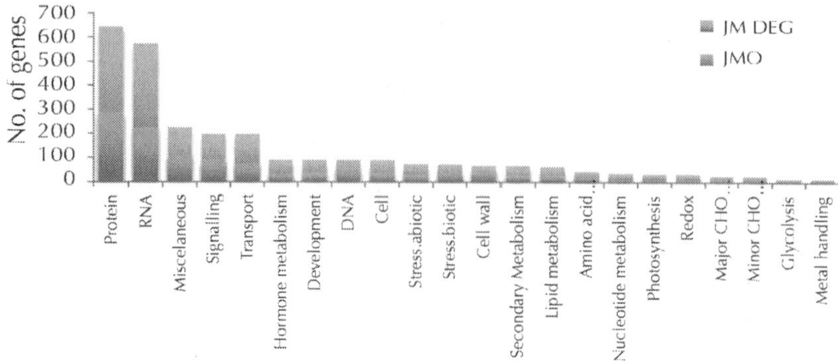

Figure 7: Functional annotations of DE genes in JM.

Only the top-level category of the MapMan annotation for each gene was considered for simplicity. Blue: all DEGs in JM; Red: JMO. Categories with at least 10 genes are shown. Genes with unknown annotations are not included in the graph.

doi:10.1371/journal.pone.0081729.g007.

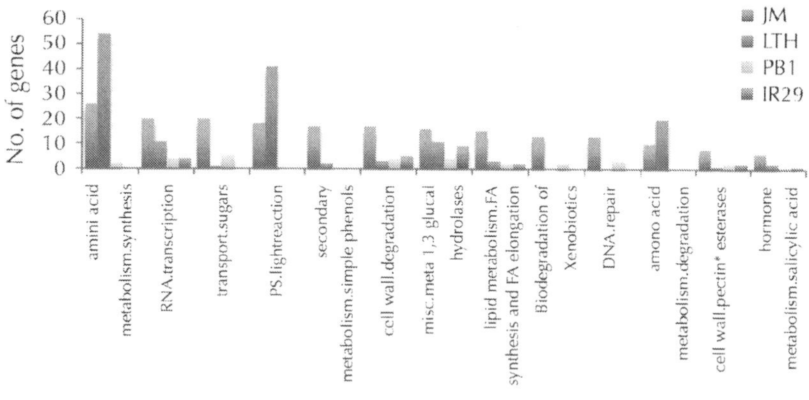

Figure 8: Highly abundant annotations in JM.

MapMan annotations that were at least five times more abundant in JM compared to PB1 and IR29, doi:10.1371/journal.pone.0081729.g008

SEVERAL TF FAMILIES ARE DE IN JM

Annotations were also downloaded from the Plant Transcription Factor database (PlantTFDB)[42] to assess the most prominent cold responsive transcription factor (TF) families. In JM, 277 (5.9%) of the DEGs belonged to 37 represented TF families. TFs from well-known cold stress associated families like AP2/ERF (34 genes), NAC (24 genes), MYB-like (21 genes), bHLH (19 genes) and WRKY (14 genes) were among the over-represented families (Figure 9, Table S1). Hierarchical clustering of the AP2/ERF TFs shows that most of these genes are either highly up- or down-regulated within 24 hours of cold stress and thus form two major clusters based on their expression (Figure 10). *OsDREB1*s are among the most highly induced genes within 30 minutes of exposure to cold.

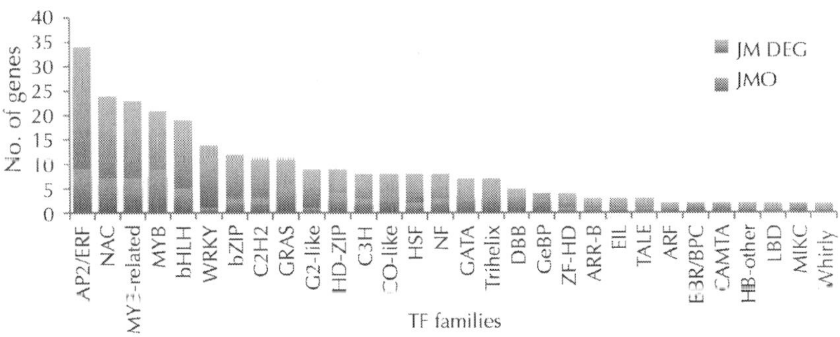

Figure 9: Differentially expressed transcription factor (TF) encoding genes, TF encoding genes were identified from the Plant Transcription Factor database (PlantTFDB). doi:10.1371/journal.pone.0081729.g009

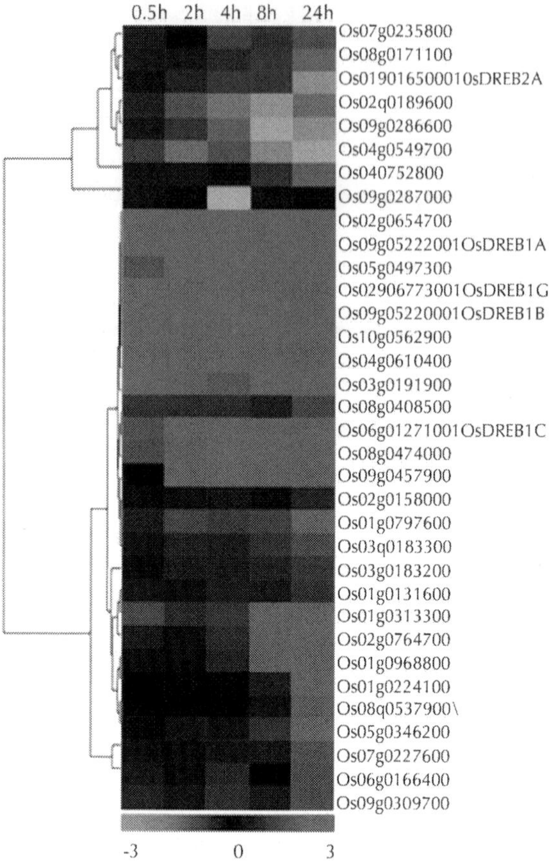

Figure 10: Clustering of AP2/ERF domain containing genes, Hierarchical clustering of differentially expressed AP2 domain containing genes. JMOs are underlined, doi:10.1371/journal.pone.0081729.g010

Significant Changes Occur in Hormone, Signaling and Carbohydrate Metabolism Associated Genes

A number of signaling, as well as hormone regulation pathways, have previously been shown to be prominent in cold stress responses [43], [44]. Genes involved in different hormone pathways were identified from the MapMan annotations. Among pathways coupled to cold stress were auxin (indole-3-acetic acid) (36 genes), abscisic acid (ABA) (15 genes), cytokinin (13 genes) and ethylene (13 genes) (Figure 11a). Genes related to calcium signaling (40 genes) and the MAP kinase cascade (8 genes) were also differentially expressed (Figure 11b), as well as genes involved in sugar synthesis and degradation pathways, including sucrose degradation (11 genes), trehalose synthesis (8 genes) and starch degradation (8 genes) (Figure 11c).

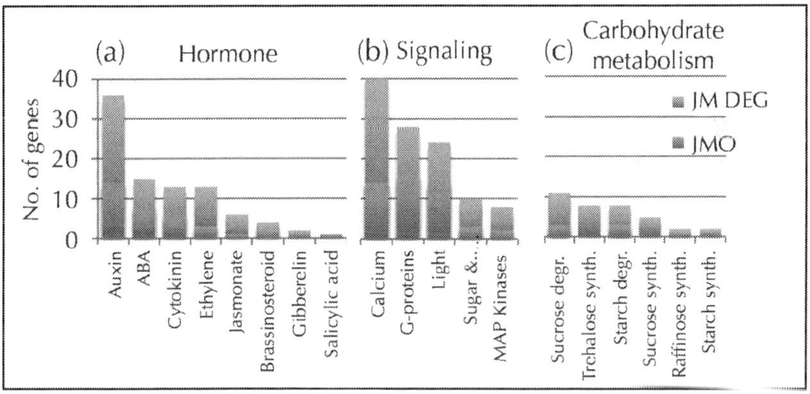

Figure 11: Genes in hormone, signaling and carbohydrate metabolism pathways, Number of cold induced genes involved in various pathways, as identified from MapMan annotations, doi:10.1371/journal.pone.0081729.g011.

COLD STRESS ACTIVATES ROS SCAVENGING MECHANISMS IN JM

Cold temperature causes oxidative stress leading to accumulation of reactive oxygen species (ROS), such as $O_2^{\cdot -}$, H_2O_2, $^1O_2$, $HO_2^{\cdot -}$, $OH^{\cdot}$, ROOH, $ROO^{\cdot}$, and $RO^{\cdot}$ [45]. ROS are highly reactive and induce deleterious effects on membranes that lead to ion leakage and also cause damage to proteins, DNA and lipids. Ultimately these effects result in cell death [45]. It has been shown previously that in the rice cv. Nipponbare, H_2O_2 levels increase within 1.5 hours of chilling stress (+10°C) [25]. To counteract the increased ROS levels, the cell produces scavengers such as peroxidases, catalases, ascorbate, glutathione, superoxide dismutase, glutaredoxins and thioredoxins that detoxify ROS [45]. In JM, 36 genes associated with ROS scavenging were cold induced. Among peroxidases, nine genes were differentially expressed, two up-regulated after eight hours and the remaining seven genes down-regulated after two to eight hours. Two glutathione peroxidases were also differentially expressed, one gene up- and one down-regulated after 24 hours, while glutathione S transferase was induced with two genes up-regulated at 8 hours and three down-regulated after four hours of cold stress. In addition, twenty genes belonging to glutaredoxins and thioredoxins were differentially expressed, with four genes up-regulated within the first two hours of cold stress and the remaining up-regulated during the 24 hour assay period. Out of the remaining 16 genes, one was up-regulated after 24 hours and the others were down-regulated (Figure 12, Table S5). Of the 36 genes, eleven genes were in the JMO set (underlined in Figure 12), one (Os01g0667900) was in the CCI set and four (Os03g0762300, Os11g0284900, Os12g0188700, Os08g0378900) were in the CIT set. Overall, this suggests differences in the ROS scavenging mechanisms in different rice cultivars.

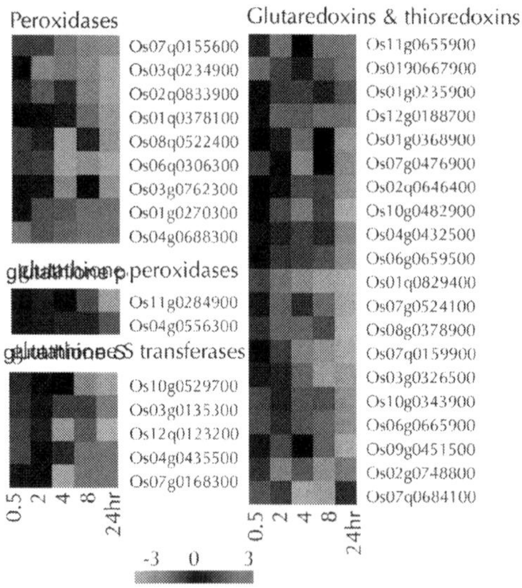

Figure 12: Clustering of genes associated with ROS scavenging, Hierarchical clustering of differentially expressed genes associated with ROS scavenging. JMOs are underlined, doi:10.1371/journal.pone.0081729.g012.

Cellular Components Play Distinct Roles under Cold Stress

Organelles and sub-cellular components are highly affected by cold stress [46]. For example, in *Arabidopsis,* 184 nuclear proteins [47], 43 chloroplast proteins [48] and 38 plasma membrane proteins were identified as differentially expressed under cold stress [49]. Mitochondria, which are the prime sites for ROS production in abiotic stress, regulate ROS levels through their energy dissipating systems [50]. To evaluate if sub-cellular components play an active role during cold stress in JM, a functional annotation analysis was performed for DEGs encoding proteins predicted to be either organelle localized or associated to a cellular structure (hereon referred to as 'orgLoc genes') (Figure 13, Table S1).

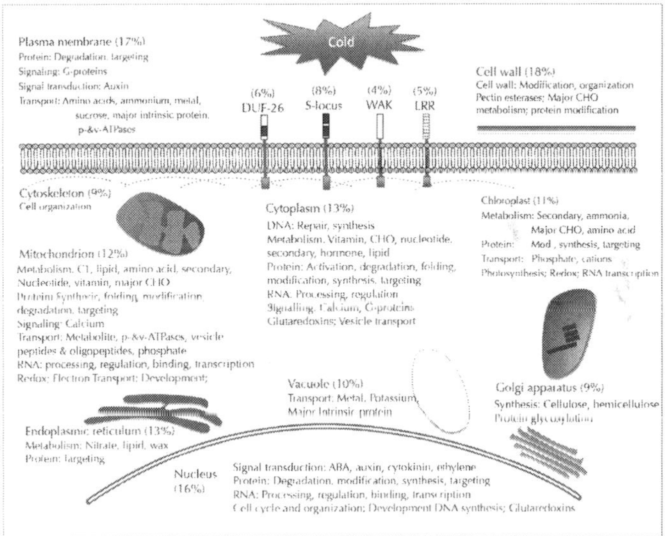

Figure 13: Overview of the role of cellular components under cold stress.

Localization of genes in cellular components was identified from Gene Ontology. Molecular functions for the corresponding genes were identified from the MapMan annotations. More detailed annotations are in Table S1.

doi:10.1371/journal.pone.0081729.g013.

The analysis showed that orgLoc genes from various cellular components were differentially expressed at a significant level during cold stress. The percentage of induced genes for each sub-cellular component or organelle was: cell wall (18%), plasma membrane (17%), nucleus (16%), cytoplasm (13%), endoplasmic reticulum (13%), mitochondria (12%), chloroplast (11%), vacuole (10%), Golgi apparatus (9%) and cytoskeleton (9%) (Figure 13). Moreover, several genes encoding plasma membrane receptors were also cold induced. A functional analysis of the orgLoc genes suggested distinct roles for different sub-cellular components and organelles during cold stress. For example, cold induced alterations in transport of metabolite process were found in the plasma membrane, mitochondria, chloroplast and vacuole, and signaling process in the plasma membrane, mitochondria and cytoplasm. Differential regulation in auxin signal transduction process could be seen in the plasma membrane, while, in the nucleus, auxin, ABA, cytokinin and ethylene signal transduction process were observed. Metabolism process of several compounds was also seen in the mitochondria, endoplasmic reticulum, cytoplasm, chloroplast and cell wall while differences in redox activity process were seen mainly in mitochondria and chloroplasts. Together this strongly suggests that the buildup of tolerance to low temperature stress involves many different components and that these components act synergistically.

IDENTIFICATION OF COLD RESPONSIVE GENES IN QTLS ASSOCIATED WITH COLD STRESS

Previous studies have identified 22 different QTLs associated with cold stress in rice (hereon referred to as 'cold QTLs'). To assess whether any of these QTLs overlap with stress response genes in JM, all 22 QTLs were downloaded from the Gramene database [51] and putative genes within the QTLs were identified and checked for homology to the JM DEGs.

This showed that 473 of the DEGs were present in 13 of the cold QTLs (Figure 14, Table S6) while no DEGs were found in the remaining 9 QTLs. MapMan annotations of the genes present in the 13 cold QTLs suggest distinct functional roles of these QTLs. The most common annotations were protein (62 genes), RNA (54 genes), signaling (27 genes), miscellaneous (22 genes) and transport (20 genes). All 13 QTLs also contained genes with yet unknown molecular functions. Of the 473 genes, 232 genes were found in the JMO set, while 40 and 12 genes were found in the CIT and CCI sets respectively (Table S7). The fact that several different QTLs contained DEGs with different functions suggests that cold tolerance is a multigenic trait. To increase cold tolerance by breeding, several different processes would, therefore, need to be successfully transferred to the same genome. As the cold QTLs were identified through genetic studies, it is highly likely that several of the cold induced genes located within the identified QTLs are critical for cold stress tolerance also in JM.

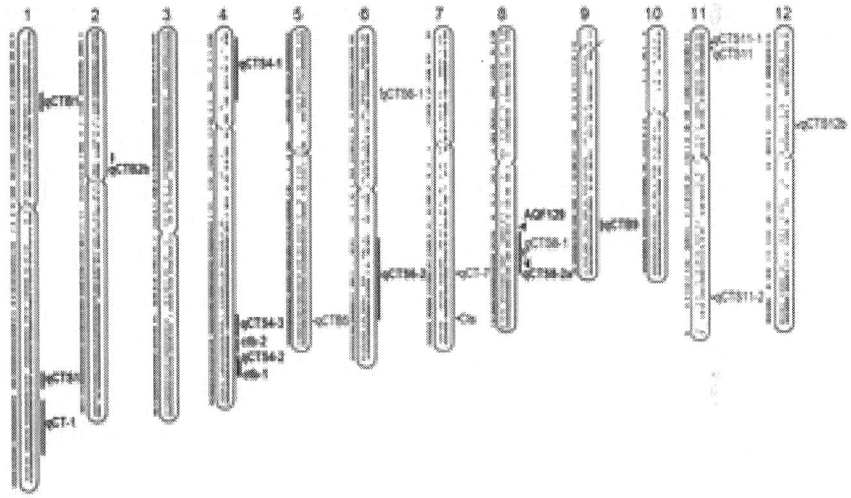

Figure 14: Localization of genes and QTLs on the chromosomes.

Chromosomal localization of the 4,636 differentially expressed genes and 22 QTLs previously identified as associated with cold stress. Red dots indicate up-regulated genes, green indicate down-regulated genes and blue dots indicate JMO. QTLs are marked with vertical lines that cover their physical ranges. Physical locations of genes and QTLs were downloaded from the Gramene database, and the figure was generated with an R script.

doi:10.1371/journal.pone.0081729.g014

DISCUSSION

Low temperatures negatively affect both productivity and yield of rice grown in colder regions of the tropics. However, over the years, farmers and breeders have selected cultivars that are more chilling tolerant, resulting in rice cultivars that can even cope with cold stress to some extent. Since we were interested in the molecular mechanisms underlying this adaptation, we selected a chilling tolerant rice variety Jumli Marshi (JM) as a model and studied its transcriptional responses to cold stress. JM is a locally adapted endogenous variety that is routinely grown at up to 3,050 m in the Jumla district of Nepal. As a consequence of the high altitude, this variety is commonly exposed to temperatures down to or below 4°C during the early developmental stages in the growing season. JM is commonly used as a parental donor in various rice breeding programs worldwide for the development of new chilling tolerant rice cultivars. Chilling tolerant Nepalese rice cultivars Manjushree-2 and Khumal-8 were developed with JM as one of the parents [52]. JM was also successfully used as a parental donor for the development of upland chilling tolerant rice FOFIFA-172 in Madagascar [53].

JM has mostly been crossed with *indica* cultivars for the development of new chilling tolerant cultivars. Rice cultivar Khumal-8 is such an example where the parents are JM (spp. *japonica*) and IR36 (spp. *indica*). A study on the pedigrees of 28 Nepalese rice cultivars showed that 65.7% of the ancestors were of *indica* type and only 2.8% were *japonica* [54] again indicating that the parental types have mostly been indica Thus, in general a more detailed molecular comparison of *japonica* and *indica* sub-types is necessary for increased understanding of physiological, molecular and evolutionary aspects of chilling tolerance in rice. A few studies where chilling tolerant *japonica* rice cultivars have been compared to chilling sensitive *indica*ones exist [25], [27], but more comparisons with different cultivars are needed. Due to their superior agronomical characteristics, *indica* cultivars are most commonly used in Nepalese rice breeding programs, while, *japonica* cultivars are valued mainly for their chilling tolerance characteristics. Thus, comparative analysis of chilling sensitive *indica* and chilling tolerant*japonica* rice will facilitate in understanding of the chilling tolerance mechanisms and identify markers for breeding. In this work, a comparative transcriptome analysis was done between chilling tolerant upland Nepalese rice JM with a chilling tolerant japonica (LTH), and the two chilling sensitive *indica* rice cultivars PB1 and IR29. Adding chilling sensitive *japonica* rice to this analysis would have been of interest, but unfortunately there is no publicly available microarray data generated from chilling sensitive japonica rice.

Many of the cold induced genes derived in this study have previously been associated to cold stress in other systems [14], [16], [17], [23], [26], [27], [55], [56]. For example, *DREB1*s act as up-stream regulators during cold stress and are induced several folds within 30 minutes of cold treatment in several plant species [14], [16], [26], [27], [55], [57], [58], [59], [60], [61],[62]. Thus, many of the early cold induced genes seen in cold hardy plants are present in rice and the overall regulation of the upstream genetic regulators is also conserved. This indicates that some of the differences between chilling and freezing tolerant crops is not in the primary activation of cold acclimation pathways but may lie in the second wave of activation of downstream protection mechanisms or the existence of CBF-independent regulons in the chilling tolerant rice cultivars. However, the OsDREB1C regulon targets [38] are cold induced in both JM and IR64 (Figure S3b), which further strengthens the hypothesis that CBF-independent regulons play a critical role in chilling tolerance in rice.

In a previous study, Yun et al. [25] identified an early response putative regulatory network in*japonica* rice upon chilling stress at 10°C and suggested a strong integration of defense and growth related responses. Several well-known cold stress responsive TFs were differentially expressed including AP2, bZIP, MYB, WRKY, bHLH and NAC families. Functional annotations such as transport, signaling, defense, transcription and apoptosis were over-represented. In this work, although JM was cold stressed at 4°C, functional annotations common under chilling stress (10°C) were also found to be common under cold stress, indicating an overlap in regulatory mechanisms in chilling and cold stress. In a transcriptome profile comparison between LTH and IR29, Zhang et al. [27] identified 1,256 genes to be up-regulated only in LTH upon cold exposure. They also identified several functional annotations specific to either LTH or to IR29 or common to both. The results from this work further highlight genes, functional annotations and regulatory motifs that are specifically induced in JM. Chlorophyll measurements suggest that IR64 plants suffer greatly within 48 hours of cold stress indicating that genes induced within the first 24 hours are critical for plant survival. This study identified 4,636 genes that were significantly differentially expressed within 24 hours upon cold stress in JM. Functional annotations of these genes showed that many of them are involved in gene regulation, post-translational modification, signaling, transport, development and metabolism. Comparative analysis with one other chilling tolerant rice and two chilling sensitive rice resulted in identification of 182 genes that were cold induced in all four rice cultivars (CCI genes). Also, 511 genes were induced in the two chilling tolerant cultivars (CIT genes). Promoter analysis of the CCI and CIT clusters gave further insight into similarities and differences among different cultivars. Motifs binding to CBF/DRE or ABRE TFs were over-represented in the promoters of CCI genes, thus suggesting that well-known cold induction pathways are conserved in rice cultivars differing in their chilling tolerance. Interestingly, ABRE elements were over-represented in the promoters of CIT genes, but CBF/DRE elements were not. This indicates that there are differences in the gene regulatory mechanisms in the chilling tolerant and sensitive lines. These differences could either be activation of parallel pathways that are independent of CBF regulation, or activation of pathways that are downstream of the CBF regulatory network. A further difference in the response of chilling tolerant and sensitive rice cultivars are the functional annotations of the DE genes. The top three annotations that were abundant in JM compared to the chilling sensitive cultivars were amino acid metabolism, RNA transcription and sugar transport.

Gene expression profiling under cold stress in *Arabidopsis* showed that the *AP2/ERF* family of TF genes was the most represented [63]. In *japonica* rice, it was shown that molecular responses triggered by oxidative signals are critical for survival under low temperature stress[25]. Under chilling stress, the chloroplast electron transport chain is over-reduced, resulting in increased production of ROS [64]. Less effective mitochondrial electron transport during cold stress can also lead to ROS formation [65], which can be detoxicated by different scavenging antioxidant enzymes produced by the plants. ROS scavengers, such as catalase [66], ascorbate peroxidase [67], glutathione reductase [68] and glutathione peroxidase [69] have been shown to play a role in eliminating free oxygen radicals during chilling stress. In rice, putative ROS mediated regulatory modules under chilling stress (+10°C) have been identified, and were suggested to be independent of DREB and ABA regulons [23]. In this work, genes involved in production of peroxidases, glutathione peroxidases, glutathione *S*-transferases, glutaredoxins and thioredoxins were significantly differentially expressed (Figure 12), indicating that ROS scavenging mechanisms are active in JM during cold stress (+4°C).

This study also shows that various cellular components play distinct roles during cold stress. Several orgLoc genes from nucleus, plasma membrane, endoplasmic reticulum and other components were differentially expressed in JM. The analysis reflects the physiological processes and downstream molecular mechanisms that possibly occur in the cellular components under cold stress. The plasma membrane acts as a primary barrier between the cytoplasm and the extracellular space and plays a critical role in cell integrity, signaling and transport [70]. In this work, plasma membrane orgLoc genes were involved in transport and the auxin signal transduction process. The cell wall provides cell integrity and support, and cell wall orgLoc genes were involved in the cell wall modification and organization process, reflecting changes occurring in the cell wall upon cold stress. The mitochondria and chloroplast are involved in metabolism of several compounds as well as energy generation from the electron transport chain. Again, several orgLoc genes with various functions differentially expressed in these cellular components indicate critical roles for these under cold stress. An overview of the processes taking place in each component during cold stress is shown in Figure 13.

Although efforts have been made to introduce increased chilling tolerance in rice by traditional breeding, the progress has been limited because chilling tolerance is a multigenic trait and involves several layers of protection to shield the plants from the stress. However, a number of QTLs that are statistically significantly associated with an increase in chilling tolerance have been identified through years of breeding. A list of such 'cold QTLs' from rice is summarized in Table S6. Since these QTLs are coupled to chilling tolerance, it is of interest to identify cold induced genes located within them. In a previous study, 445 cold induced genes from rice were located to 21 QTLs [27]. In this work, 473 of the cold induced JM genes were found to be located within 13 of the 22 previously identified cold QTLs (Figure 14, Table S6). Of these 473 genes, 232 genes were found in the JMO set, while

40 and 12 genes were found in the CIT and CCI sets respectively (Table S7). A more substantial increase in chilling tolerance is most likely only possible by a combined contribution from multiple QTLs. Thus, this and other similar studies on QTLs will facilitate the development of new markers and will further our understanding of chilling tolerance mechanisms in rice.

CONCLUSION

Jumli Marshi is a highly adapted upland chilling tolerant Nepalese rice variety and is used extensively in breeding for new chilling tolerant rice cultivars both in Nepal and other countries. In this study, a global expression profiling was done to identify cold induced genes in JM. Comparative analysis resulted in identification of 2,101 genes DE only in chilling tolerant JM rice. From the comparison of JM and LTH, it can be concluded that there are significant differences in the way the two chilling tolerant japonica rice cultivars respond to cold, which further emphasizes the importance of including locally adapted landraces in scientific and breeding programs. In addition, in JM, 473 cold induced genes localized in previously described cold associated QTLs were identified and among these, 232 genes were DE only in JM. Further validation of these genes may lead to the development of new molecular markers for breeding of rice cultivars with increased chilling tolerance.

REFERENCES

1. Fowler DB, Carles RJ (1979) Growth, Development, and Cold Tolerance of Fall-acclimated Cereal Grains. Crop Science 19: 915–922. doi: 10.2135/cropsci1979.0011183x001900060040x
2. Thomashow MF (1999) PLANT COLD ACCLIMATION: Freezing Tolerance Genes and Regulatory Mechanisms. Annu Rev Plant Physiol Plant Mol Biol 50: 571–599. doi: 10.1146/annurev.arplant.50.1.571
3. Anderson MD, Prasad TK, Martin BA, Stewart CR (1994) Differential Gene Expression in Chilling-Acclimated Maize

Seedlings and Evidence for the Involvement of Abscisic Acid in Chilling Tolerance. Plant Physiol 105: 331–339.

4. Gesch RW, Heilman JL (1999) Responses of photosynthesis and phosphorylation of the light-harvesting complex of photosystem II to chilling temperature in ecologically divergent cultivars of rice. Environmental and Experimental Botany 41: 257–266. doi: 10.1016/s0098-8472(99)00008-8
5. Howell KA, Narsai R, Carroll A, Ivanova A, Lohse M, et al. (2009) Mapping metabolic and transcript temporal switches during germination in rice highlights specific transcription factors and the role of RNA instability in the germination process. Plant physiology 149: 961–980. doi: 10.1104/pp.108.129874
6. Morsy MR, Jouve L, Hausman JF, Hoffmann L, Stewart JM (2007) Alteration of oxidative and carbohydrate metabolism under abiotic stress in two rice (Oryza sativa L.) genotypes contrasting in chilling tolerance. Journal of plant physiology 164: 157–167. doi: 10.1016/j.jplph.2005.12.004
7. Chawade A, Linden P, Brautigam M, Jonsson R, Jonsson A, et al. (2012) Development of a model system to identify differences in spring and winter oat. PloS one 7: e29792. doi: 10.1371/journal.pone.0029792
8. Laudencia-Chingcuanco D, Ganeshan S, You F, Fowler B, Chibbar R, et al. (2011) Genome-wide gene expression analysis supports a developmental model of low temperature tolerance gene regulation in wheat (Triticum aestivum L.). BMC genomics 12: 299. doi: 10.1186/1471-2164-12-299
9. Svensson JT, Crosatti C, Campoli C, Bassi R, Stanca AM, et al. (2006) Transcriptome analysis of cold acclimation in barley albina and xantha mutants. Plant physiology 141: 257–270. doi: 10.1104/pp.105.072645
10. Janska A, Aprile A, Zamecnik J, Cattivelli L, Ovesna J (2011) Transcriptional responses of winter barley to cold indicate nucleosome remodelling as a specific feature of crown tissues. Functional & integrative genomics 11: 307–325. doi: 10.1007/s10142-011-0213-8
11. Bräutigam M, Lindlöf A, Zakhrabekova S, Gharti-Chhetri G, Olsson B, et al. (2005) Generation and analysis of 9792 EST

sequences from cold acclimated oat, Avena sativa. BMC Plant Biology 5: 18. doi: 10.1186/1471-2229-5-18

12. Kilian J, Whitehead D, Horak J, Wanke D, Weinl S, et al. (2007) The AtGenExpress global stress expression data set: protocols, evaluation and model data analysis of UV-B light, drought and cold stress responses. Plant J 50: 347–363. doi: 10.1111/j.1365-313x.2007.03052.x
13. Kreps JA, Wu Y, Chang HS, Zhu T, Wang X, et al. (2002) Transcriptome changes for Arabidopsis in response to salt, osmotic, and cold stress. Plant Physiol 130: 2129–2141. doi: 10.1104/pp.008532
14. Fowler S, Thomashow MF (2002) Arabidopsis transcriptome profiling indicates that multiple regulatory pathways are activated during cold acclimation in addition to the CBF cold response pathway. Plant Cell 14: 1675–1690. doi: 10.1105/tpc.003483
15. Benedict C, Geisler M, Trygg J, Huner N, Hurry V (2006) Consensus by democracy. Using meta-analyses of microarray and genomic data to model the cold acclimation signaling pathway in Arabidopsis. Plant Physiol 141: 1219–1232. doi: 10.1104/pp.106.083527
16. Maruyama K, Sakuma Y, Kasuga M, Ito Y, Seki M, et al. (2004) Identification of cold-inducible downstream genes of the Arabidopsis DREB1A/CBF3 transcriptional factor using two microarray systems. Plant J 38: 982–993. doi: 10.1111/j.1365-313x.2004.02100.x
17. Seki M, Narusaka M, Ishida J, Nanjo T, Fujita M, et al. (2002) Monitoring the expression profiles of 7000 Arabidopsis genes under drought, cold and high-salinity stresses using a full-length cDNA microarray. Plant J 31: 279–292. doi: 10.1046/j.1365-313x.2002.01359.x
18. Lindlof A, Brautigam M, Chawade A, Olsson O, Olsson B (2009) In silico analysis of promoter regions from cold-induced genes in rice (Oryza sativa L.) and Arabidopsis thaliana reveals the importance of combinatorial control. Bioinformatics 25: 1345–1348. doi: 10.1093/bioinformatics/btp172
19. Chawade A, Brautigam M, Lindlof A, Olsson O, Olsson B (2007) Putative cold acclimation pathways in Arabidopsis thaliana identified by a combined analysis of mRNA co-expression

patterns, promoter motifs and transcription factors. BMC genomics 8: 304. doi: 10.1186/1471-2164-8-304

20. Cooper B, Clarke JD, Budworth P, Kreps J, Hutchison D, et al. (2003) A network of rice genes associated with stress response and seed development. Proceedings of the National Academy of Sciences of the United States of America 100: 4945–4950. doi: 10.1073/pnas.0737574100
21. Lindlöf A, Bräutigam M, Chawade A, Olsson B, Olsson O (2007) Identification of Cold-Induced Genes in Cereal Crops and Arabidopsis Through Comparative Analysis of Multiple EST Sets. 4414: 48–65. doi: 10.1007/978-3-540-71233-6_5
22. Rabbani MA, Maruyama K, Abe H, Khan MA, Katsura K, et al. (2003) Monitoring expression profiles of rice genes under cold, drought, and high-salinity stresses and abscisic acid application using cDNA microarray and RNA gel-blot analyses. Plant physiology 133: 1755–1767. doi: 10.1104/pp.103.025742
23. Cheng C, Yun KY, Ressom HW, Mohanty B, Bajic VB, et al. (2007) An early response regulatory cluster induced by low temperature and hydrogen peroxide in seedlings of chilling-tolerant japonica rice. BMC genomics 8: 175. doi: 10.1186/1471-2164-8-175
24. Oda S, Kaneko F, Yano K, Fujioka T, Masuko H, et al. (2010) Morphological and gene expression analysis under cool temperature conditions in rice anther development. Genes & genetic systems 85: 107–120. doi: 10.1266/ggs.85.107
25. Yun KY, Park MR, Mohanty B, Herath V, Xu F, et al. (2010) Transcriptional regulatory network triggered by oxidative signals configures the early response mechanisms of japonica rice to chilling stress. BMC plant biology 10: 16. doi: 10.1186/1471-2229-10-16
26. Mittal D, Madhyastha DA, Grover A (2012) Genome-wide transcriptional profiles during temperature and oxidative stress reveal coordinated expression patterns and overlapping regulons in rice. PloS one 7: e40899. doi: 10.1371/journal.pone.0040899
27. Zhang T, Zhao X, Wang W, Pan Y, Huang L, et al. (2012) Comparative transcriptome profiling of chilling stress responsiveness in two contrasting rice genotypes. PloS one 7: e43274. doi: 10.1371/journal.pone.0043274

28. Sapkota S, Paudel MN, Thakur NS, Nepali MB, Neupane R (2010) Effect of Climate Change on Rice Production: A Case of Six VDCs in Jumla District. Nepal Journal of Science and Technology 11: 57–62. doi: 10.3126/njst.v11i0.4124
29. Caldana C, Scheible WR, Mueller-Roeber B, Ruzicic S (2007) A quantitative RT-PCR platform for high-throughput expression profiling of 2500 rice transcription factors. Plant Methods 3: 7. doi: 10.1186/1746-4811-3-7
30. Su CF, Wang YC, Hsieh TH, Lu CA, Tseng TH, et al. (2010) A novel MYBS3-dependent pathway confers cold tolerance in rice. Plant physiology 153: 145–158. doi: 10.1104/pp.110.153015
31. Pfaffl MW (2001) A new mathematical model for relative quantification in real-time RT-PCR. Nucleic acids research 29: e45. doi: 10.1093/nar/29.9.e45
32. Gentleman RC, Carey VJ, Bates DM, Bolstad B, Dettling M, et al. (2004) Bioconductor: open software development for computational biology and bioinformatics. Genome biology 5: R80.
33. Bolstad BM (2004) Low Level Analysis of High-density Oligonucleotide Array Data: Background, Normalization and Summarization. Berkeley: University of California.
34. Gautier L, Cope L, Bolstad BM, Irizarry RA (2004) affy–analysis of Affymetrix GeneChip data at the probe level. Bioinformatics 20: 307–315. doi: 10.1093/bioinformatics/btg405
35. Smyth GK (2005) Limma: linear models for microarray data. In: Gentleman R, Carey V, Dudoit S, Irizarry R, Huber W, editors. Bioinformatics and computational biology solutions using R and Bioconductor. New York: Springer Science+Business Media. pp. 397–420.
36. Ernst J, Bar-Joseph Z (2006) STEM: a tool for the analysis of short time series gene expression data. BMC bioinformatics 7: 191. doi: 10.1093/bioinformatics/bti1022
37. Gilmour SJ, Fowler SG, Thomashow MF (2004) Arabidopsis transcriptional activators CBF1, CBF2, and CBF3 have matching functional activities. Plant molecular biology 54: 767–781. doi: 10.1023/b:plan.0000040902.06881.d4

38. Zhang FD, Huang LM, Wang WD, Zhao XD, Zhu LM, et al. (2012) Genome-wide gene expression profiling of introgressed indica rice alleles associated with seedling cold tolerance improvement in a japonica rice background. BMC genomics 13: 461. doi: 10.1186/z
39. Siddiq EA, Vemireddy LR, Nagaraju J (2012) Basmati Rices: Genetics, Breeding and Trade. Agricultural Research 1: 25–36. doi: 10.1007/s40003-011-0011-5
40. Morris RT, O'Connor TR, Wyrick JJ (2008) Osiris: an integrated promoter database for Oryza sativa L. Bioinformatics. 24: 2915–2917. doi: 10.1093/bioinformatics/btn537
41. Mockler TC, Michael TP, Priest HD, Shen R, Sullivan CM, et al. (2007) The DIURNAL project: DIURNAL and circadian expression profiling, model-based pattern matching, and promoter analysis. Cold Spring Harb Symp Quant Biol 72: 353–363. doi: 10.1101/sqb.2007.72.006
42. Perez-Rodriguez P, Riano-Pachon DM, Correa LG, Rensing SA, Kersten B, et al. (2010) PlnTFDB: updated content and new features of the plant transcription factor database. Nucleic acids research 38: D822–827. doi: 10.1093/nar/gkp805
43. Huang GT, Ma SL, Bai LP, Zhang L, Ma H, et al. (2012) Signal transduction during cold, salt, and drought stresses in plants. Molecular biology reports 39: 969–987. doi: 10.1007/s11033-011-0823-1
44. Xiong L, Schumaker KS, Zhu JK (2002) Cell signaling during cold, drought, and salt stress. The Plant cell 14 Suppl: S165–183
45. Gill SS, Tuteja N (2010) Reactive oxygen species and antioxidant machinery in abiotic stress tolerance in crop plants. Plant physiology and biochemistry : PPB/Societe francaise de physiologie vegetale 48: 909–930. doi: 10.1016/j.plaphy.2010.08.016
46. Hossain Z, Nouri MZ, Komatsu S (2012) Plant cell organelle proteomics in response to abiotic stress. J Proteome Res 11: 37–48. doi: 10.1021/pr200863r
47. Bae MS, Cho EJ, Choi EY, Park OK (2003) Analysis of the Arabidopsis nuclear proteome and its response to cold stress. The Plant journal : for cell and molecular biology 36: 652–663. doi: 10.1046/j.1365-313x.2003.01907.x

67. Wang Y, Wisniewski M, Meilan R, Cui M, Webb R, et al. (2005) Over- expression of cytosolic ascorbate peroxidase in tomato confers tolerance to chilling and salt stress. J Am Soc Hortic Sci 130: 167–173.

68. Kornyeyev D, Logan BA, Payton PR, Allen RD, Holaday AS (2003) Elevated chloroplastic glutathione reductase activities decrease chilling-induced photoinhibition by increasing rates of photochemistry, but not thermal energy dissipation, in transgenic cotton. Functional Plant Biology 30: 101. doi: 10.1071/fp02144

69. Yoshimura K, Miyao K, Gaber A, Takeda T, Kanaboshi H, et al. (2004) Enhancement of stress tolerance in transgenic tobacco plants overexpressing Chlamydomonas glutathione peroxidase in chloroplasts or cytosol. The Plant journal : for cell and molecular biology 37: 21–33. doi: 10.1046/j.1365-313x.2003.01930.x

70. Dowgert MF, Steponkus PL (1984) Behavior of the Plasma Membrane of Isolated Protoplasts during a Freeze-Thaw Cycle. Plant physiology 75: 1139–1151. doi: 10.1104/pp.75.4.1139

71. Oliveros JC (2007) VENNY. An interactive tool for comparing lists with Venn Diagrams.

Chapter 4

Genes Controlling Root Development in Rice

Chung D Mai[1,2], Nhung TP Phung[1,3,4], Huong TM To[2], Mathieu Gonin[3], Giang T Hoang[1,2], Khanh L Nguyen[2,3], Vinh N Do[1], Brigitte Courtois[4] and Pascal Gantet[2,3,5*]

[1]Agricultural Genetic Institute, LMI RICE, Hanoi, Vietnam

[2]University of Science and Technology of Hanoi, LMI RICE, Hanoi, Vietnam

[3]IRD, UMR DIADE, LMI RICE, Hanoi, Vietnam

[4]CIRAD, UMR AGAP, Montpellier, France

[5]Université Montpellier 2, UMR DIADE, Montpellier, France

ABSTRACT

In this review, we report on the recent developments made using both genetics and functional genomics approaches in the discovery of genes controlling root development in rice. QTL detection in classical biparental mapping populations initially enabled the identification of a very large number of large chromosomal segments carrying root genes. Two segments with large effects have been positionally cloned, allowing the identification of two major genes. One of these genes conferred a tolerance to low phosphate content in soil, while the other conferred a tolerance to drought by controlling root gravitropism, resulting in root system expansion deep in the soil. Findings based on the higher-resolution QTL detection offered by the development of association mapping are discussed. In parallel with genetics approaches, efforts have been made to screen mutant libraries for lines presenting alterations in root development, allowing for the identification of several genes that control different steps of root development, such as crown root and lateral root initiation and emergence, meristem patterning, and the control of root growth. Some of these genes are closely phylogenetically related to *Arabidopsis* genes involved in the control of lateral root initiation. This close relationship stresses the conservation among plant species of an auxin responsive core gene regulatory network involved in the control of post-embryonic root initiation. In addition, we report on several genetic regulatory pathways that have been described only in rice. The complementarities and the expected convergence of the direct and reverse genetic approaches used to decipher the genetic determinants of root development in rice are discussed in regards to the high diversity characterizing this species and to the adaptations of rice root system architecture to different edaphic environments.

INTRODUCTION

Roots are essential organs for exploring and exploiting soil resources, such as water and mineral nutrients. Different root architecture ideotypes that are adapted to different soil mineral nutrient balances or water statuses have been proposed. It is generally acknowledged that a deeper, thicker and more branched root system with a high root to shoot ratio can enhance the tolerance of rice to water deficits (Fukai and Cooper, [1995]; Gowda et al. [2011]). Roots develop primarily underground and are inaccessible to direct observation. This limitation partly explains why root architecture is not a widespread breeding objective. Because of this lack of past investment, there is room for improvement of rice root systems. This task is particularly important in a context of climate change, which will likely increase the occurrence of unfavorable conditions (Den Herder et al. [2010]). An improved root system can be obtained more easily using indirect selection methods, such as marker-assisted selection. It is therefore necessary to identify the main genetic determinants governing root development. Root development is a complex process that involves constitutive and adaptive mechanisms and regulatory correlations with the shoot part of the plant (Puig et al., [2012]). In this review, we focus on genetic determinants of root development in rice that were identified using complementary direct and reverse genetic approaches, such as QTL detection, mutant analysis and transcriptomics. The involvement of these genes in plant adaptation to stress conditions is discussed when such information is available and relevant.

REVIEW

Main Characteristics of the Rice Root System

The rice root system is composed of a seminal root and postembryonic shoot borne-roots termed crown roots (Rebouillat et al. [2009]; Coudert et al. [2010]). These two types of roots can branch and form long lateral roots or short lateral roots. Crown roots differentiate in the stem from a radial ground meristem that has common characteristics with the root pericycle (Coudert et al.[2013b]; Itoh et al. [2005]). Lateral roots differentiate from the root pericycle and, in part, from the endoderm (Rebouillat et al. [2009]; Orman-Ligeza et al. [2013]). The radial structure of the rice roots comprises the following tissues, from the center to periphery: the stele, including phloem and xylem vessels and the pericycle; the endoderm; the cortex, whose cells can undergo apoptosis to constitute the aerenchyma; the sclerenchyma; the exodermis; and the epidermis (Rebouillat et al. [2009]). This radial structure reflects the capacity of rice roots to grow in aerobic as well as anaerobic conditions. Notably, the aerenchyma enables gas exchange with the shoot when the plant is growing under anaerobic conditions. Rice was domesticated approximately 10 000 years ago (Sweeney and McCouch [2007]) and has since adapted to a large variety of ecosystems (irrigated, rainfed lowland, upland, flood-prone, and mangrove) and production systems (from traditional low-input systems to intensive high-input systems). More than 100 000 rice genotypes are available in the Genebank of the International Rice Research Institute (IRRI), Philippines. The phenotypic evaluation of a small sample of this diversity showed that the tested genotypes exhibited a great deal of root architecture variation (O'Toole and Bland [1987]; Lafitte et al. [2001]). The remaining unexplored extensive variability constitutes a very significant opportunity for the identification of new genes involved in root development.

Detection of QTLS for Root Development in Rice

To identify genes controlling root system architecture, the direct genetic approach has historically preceded the reverse genetic approach. The direct genetic approach was first based on QTL detection in mapping populations. The first study on root architecture in rice was undertaken in 1995 (Champoux et al. [1995]), and a large number of studies followed. These studies were reviewed by Kamoshita et al. ([2008]) for those concerning the CT9993/IR622266 population and[Khowaja et al. (2009)] for those concerning Bala/Azucena. Courtois et al. ([2009]) analyzed the conditions of detection of 675 QTLs in 24 studies conducted on 12 different populations. Simple phenotyping systems, generally based on hydroponic culture conditions, pots or soil columns, characterized this period of work. In the majority of the cases (368 among the 675 QTLs), these phenotyping conditions were favorable in order to estimate the crop's genetic potential; only a few studies specifically analyzed response QTLs in conditions of water stress or root penetration constraints (147 and 160 QTLs, respectively). The measured traits were those that were the most easily observable, such as maximum root length, root thickness, root number, mass of roots at different depths, or root to shoot ratio. The preferred mapping populations were recombinant inbred lines (RILs) or doubled haploid populations because the fixed nature of these lines enabled replicate measurements. To maximize the phenotypic differences between parents and simplify the identification of polymorphisms, most of the populations (8 populations representing 560 QTLs) were derived from indica x japonica crosses. The mapping population sizes were generally relatively small (between 100 and 150 accessions), enabling the identification of several QTLs with large effects but offering little resolution. Although large QTL x environment interactions were recorded (Kamoshita et al. [2002]; MacMillan et al. [2006]), some chromosomal regions appeared to be detected regularly in various populations and/or environment, as shown by meta QTL analyses (Courtois et al. [2009]; [Khowaja et al. 2009]). For the traits for which the highest number of QTL were detected, QTL clusters were found on chromosomes 1 (30–40 Mb), 2 (25–35 Mb), 3 (0–5 Mb), 4 (30–35 Mb), and 9 (15–20 Mb), as shown in Figure 1. More recently, the progress on high-throughput root phenotyping and 3D image reconstruction has led to a renewed interest in QTL detection in mapping populations, focusing on original traits (e.g., the root network bushiness or its convex hull volume), in addition to more classical traits, such as specific root length or specific root area (Topp et al. [2013]). This last study using the Bala/Azucena population revealed QTLs that had already been detected for simpler traits and identified new QTLs as well.

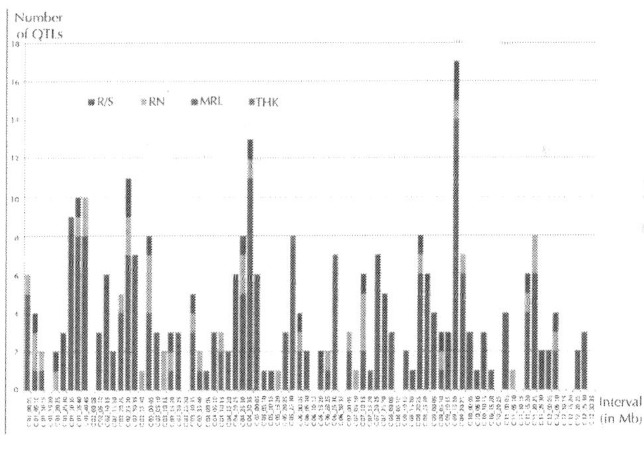

Figure 1: Number of root related QTLs detected in rice per chromosome interval.

QTL for root to shoot ratio (R/S, violet), root number (RN, green) maximum root length (MRL, red), and root thickness (THK, blue). The figure was built from data reported in Courtois et al. ([2009]).

The main limitation with QTL detection in these mapping populations is the large size of the confidence interval. Even meta-analyses, on average, did not enable a decrease in the meta-QTL confidence intervals to less than half the original size of the segment (Courtois et al. [2009]). Genome-wide association studies (GWAS) in natural populations showing fast linkage disequilibrium decay have recently appeared as powerful tools to improve the resolution in localizing QTLs compared to mapping populations (Zhu et al. [2008]). GWAS requires a very high marker density, which is permitted by the simultaneous development of next-generation sequencing technology. The first results of GWAS analyzing root traits in rice have recently been published (Clark et al. [2013]; Courtois et al. [2013]). Because of the strong bipolar structure of*O. sativa*, which induces high risk of false-positive associations, the analyses were conducted on panels composed of accessions belonging to a single varietal group. This type of panel gives access to new varietal group-specific QTLs that cannot be detected by classical mapping in indica x japonica populations. However, even if the resolution on QTL position is clearly improved in GWAS compared with mapping populations, the number of possible candidate genes is still large and requires complementary evidence to determine the functional gene(s).

Positional Cloning of Genes Underlying Root QTLS

The results of QTL mapping were pursued through the development of NILs introgressed with QTLs in different backgrounds and QTL fine mapping with the aim of positional cloning (Shen et al.[2001]; Steele et al. [2013]; Suji et al. [2012]; Steele et al. [2006]). The most interesting QTLs for which the importance of the phenotypic effect was confirmed were positionally cloned. The first of these was *PHOSPHORUS UPTAKE 1* (*PUP1*), a QTL contributing to phosphorus (P) uptake in low P content soils. The gene underlying the QTL, later termed *PHOSPHORUS-STARVATION TOLERANCE 1*(*PSTOL1*), was cloned and appeared to encode a receptor-like cytoplasmic kinase (Gamuyao et al.[2012]). The gene does not exist in Nipponbare. When overexpressed, the gene significantly increases the plant P content and enhances grain yield by more than 60% under P-deficient conditions. The gene acts by enhancing early root growth, conferring a globally larger root system to the plant, with enhanced root length and root surface area, contributing to an increased uptake of P, and of other nutrients (N and K). The gene is expressed in the zone of crown root initiation and crown root meristems in stem bases. This localization suggests that the gene is involved in the regulation of crown root formation and elongation.

Another cloned gene related to a root development QTL is *DEEPER ROOTING 1* (*DRO1*), which controls root growth angle and enhances deep rooting (Uga et al. [2013]). *DRO1* encodes an unknown protein that is associated with plasmalemma. *DRO1* expression is regulated by auxin via AUXIN RESPONSE FACTOR (ARF) transcription factors. DRO1 modulates root gravitropic response, likely via a modulation of epidermal cell elongation that enables roots to orientate their growth relative to the pull of gravity. The expression of *DRO1* in the IR64 background increases the angle between roots and the horizontal axis, inducing deeper rooting. Compared to IR64, the near isogenic lines carrying *DRO1* showed less deleterious effects, such as leaf wilting or delayed flowering, under moderate to severe drought stresses and better yield under stress, with no yield penalty under no stress conditions.

The cloning of these two QTLs that directly or indirectly affect root architecture exemplifies the difficulties that can occur in QTL cloning and can partly explain why the number of cloned root QTLs is still low. The first difficulty, which is intrinsic to the cloning process whatever the trait, is the lack of precision in positioning QTLs in mapping populations or, to a much lesser extent, in association panels. Therefore, lengthy operations are needed to reduce the size of the intervals to the gene level. A second difficulty in gene cloning is due to the importance of structural differences between varietal groups (Schatz et al. [2014]) and, within varietal groups, between varieties, as is illustrated by the absence of genes such as *PSTOL1* in the reference cultivar Nipponbare. A final reason is that many genes do not have an assigned function. For example, the *DRO1* product has no similarity to known proteins (Uga et al. [2013]). For these reasons, it is essential to complement these genetic studies that exploit the rice diversity using functional analysis approaches.

The sequencing and annotation of the rice genome (Itoh et al. [2007]; Yu et al. [2002]) as well as the creation of mutant collections (Hirochika et al. [2004]; Krishnan et al. [2009]; Wei et al.[2013]; Lorieux et al. [2012]; Guiderdoni and Gantet [2012]) and their screening for altered root phenotypes allowed the identification of major genes regulating the initiation, emergence or development of rice roots (Rebouillat et al. [2009]; Coudert et al. [2010]; Orman-Ligeza et al.[2013]). The gene networks involved in rice root development discussed below are schematized in Figure 2.

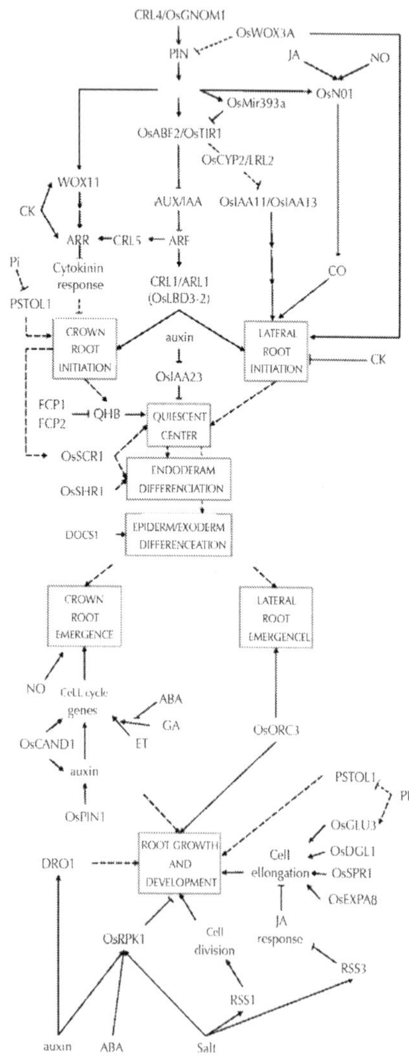

Figure 2: Gene networks controlling root development in rice .Arrows represent the positive regulatory action of one element of the network on another

one. A line ending with a trait represents the negative regulatory action of one element of the network on another one. Dotted line represent hypothetical link between two elements. Text color code: genes, black; hormones, blue; signals, red; biological processes, green.

A Conserved Core Gene Regulatory Network Regulates Post-Embryonic Root Initiation in Response to Auxin

The *crown root less 4* (*crl4*) mutant is impaired in *OsGNOM1*, which encodes a membrane-associated guanine-nucleotide exchange factor for the ADP-ribosylation factor G protein (Kitomi et al. [2008]; Liu et al. [2009]). This latter factor is a homolog of GNOM1, which regulates the intracellular traffic of the PINFORMED1 (PIN1) auxin efflux carrier protein in *Arabidopsis*. In turn, PIN1 is necessary to concentrate auxin in lateral root funder cells (Kitomi et al. [2008]; Liu et al.[2009]; Steinmann et al. [1999]). *Crl4* and *Osgnom1* mutants are characterized by an absence of crown roots and a reduced number of lateral roots. In these mutants, the expression of several*OsPIN* genes is altered. In the *crl4* mutant, auxin transport is impaired, causing auxin distribution to be modified in the stem bases (Kitomi et al. [2008]). This result suggests that local auxin transport is involved in the initiation of crown and lateral roots in rice. WUSCHEL-related homeobox 3A (OsWOX3A) is encoded by *NARROW LEAF2* (*NAL2*) and *NAL3*, a pair of duplicated genes. The *nal2/nal3* double mutant shows a complex phenotype, with altered development of different parts of the plant and, notably, a strong reduction of the density of lateral roots. This latter effect is associated with an increase in root hair number and length (Cho et al. [2013]). In*nal2/nal3* mutants, the expression of different *OsPIN* genes is modified. It is hypothesized that this modification results in a preferential allocation of auxin to the epidermis, where root hair formation is promoted. This occurs to the detriment of the pericycle, where lateral roots initiate (Yoo et al.[2013]). *OsTIR1* and *OsAFB2*, the rice orthologs of the *Arabidopsis* auxin receptors TRANSPORT INHIBITOR RESPONSE 1(TIR1) and AUXIN SIGNALING F-BOX2 (AFB2), are negatively regulated by *OsMir393a* and *OsMir393b* (Bian et al. [2012]; Xia et al. [2012]). Plants overexpressing*OsMir393* exhibit a pleiotropic altered auxin signaling phenotype, including a reduced number of crown roots. *OsMir393a* is expressed in the initiation sites of crown and lateral roots, suggesting that *OsMir393a* is involved in the regulation of post-embryonic root formation via a negative regulation of the rice auxin receptors. Interestingly, the TIR1 and AFB2 auxin receptors are localized to the nucleus, where they can interact with OsIAA1, an AUXIN/INDOLE ACETIC ACID (AUX/IAA) regulatory protein (Bian et al. [2012]; Xia et al. [2012]). AUX/IAA proteins interact with and inhibit the auxin response factors (ARF) that function as transcription factors. In *Arabidopsis*, when the TIR1 receptor is complexed with auxin, it can interact with AUX/IAA proteins. This binding results in the recruitment of a SKP1-CULLIN-FBOX (SCF) ubiquitin ligase, which ubiquitinates the AUX/IAA protein, targeting it for degradation. These interactions require other chaperone and co-chaperone proteins (Guilfoyle and Hagen [2007]; Vanneste and Friml [2009]; Mockaitis and Estelle [2008]; Lavenus et al. [2013]). In rice, a mutant screen for a loss of auxin sensitivity resulted in the characterization of two mutants, *Oscyclophilin2* (*Oscyp2*) and *lateral rootless2* (*lrl2*), which are characterized by an absence of lateral roots and are caused by mutations in the same gene (Kang et al. [2013]; Zheng et al. [2013]). In addition, *Oscyp2* exhibits a pleiotropic auxin-insensitive phenotype that correlates with a reduced degradation of OsIAA11 in response to auxin (Kang et al. [2013]). OsCYP2/LRL2 is a chaperone protein that can interact with the co-chaperone SUPPRESSOR OF G2 ALLELE OF SKP1 (OsSGT1). It has been proposed that this complex is involved in the degradation of OsIAA11 by SCF^{TIR} in response to auxin during the early step of lateral root initiation (Kang et al. [2013]). OsIAA11 or OsIAA13 gain of function mutations, in which amino acids involved in degradation of the protein in response to auxin are modified, exhibit a lateral rootless phenotype (Kitomi et al. [2012]; Zhu et al. [2012]). OsIAA11 and OsIAA13 are required for lateral root initiation. The *crown root less 1* (Inukai et al. [2005]) and *adventitious less 1* (Liu et al. [2005]) mutants are both impaired in the *OsLBD3-2* gene, which encodes a LATERAL ORGAN BOUNDARY DOMAIN (LBD) transcription factor belonging to a specific clade regrouping LBD proteins involved in the control of post-embryonic root initiation in different species (Coudert et al. [2013a]). *Crl1* and *arl1* mutants are characterized by a total absence of crown root initiation and a reduced number of lateral roots. The expression of *CRL1* is induced by auxin. This induction is lost in plants overexpressing a modified OsIAA3 protein that cannot be degraded after auxin treatment, showing that the induction of the expression of *CRL1* by auxin is dependent on the degradation of AUX/IAA proteins. In addition, the *CRL1* promoter contains an auxin response sequence (*ARE*) that interacts *in vitro* with the OsARF16 transcription factor (Inukai et al. [2005]). Taken together, these data show that the initiation of crown and lateral roots in rice involves a core auxin regulated pathway that includes TIR1/AFB2 auxin receptors, AUX/IAA regulatory proteins, as well as specific ARF and LBD transcription factors. This gene regulatory network that controls post-embryonic root development is well conserved in different dicot and monocot plant species (Peret et al. [2009]; Lavenus et al. [2013]; Hochholdinger and Tuberosa[2009]). In rice, some elements of this regulatory network are common for the initiation of crown and lateral roots, whereas others are specific to the initiation of only one root type. For this reason, rice constitutes a promising model to investigate how this auxin-dependent gene regulatory network has evolved to control post-embryonic root development in the context of both stems and roots (Coudert et al. [2013b]).

Post-Embryonic Root Initiation in Rice involves Genes that Mediate other Hormone Signals

The *crown root less5* (*crl5*) mutant exhibits a reduced number of crown roots due to a reduction in crown root initiation (Kitomi et al. [2011]). *CRL5* encodes an APETALA/ETHYLENE RESPONSE FACTOR (AP2/ERF) transcription factor. Its expression is induced by auxin via an ARF transcription factor. Cytokinins negatively regulate the initiation of crown and lateral roots in rice (Rani Debi et al. [2005]; Kitomi et al. [2011]), but plants that constitutively express CRL5 can initiate crown roots normally. CRL5 activates the expression of two type-A *RESPONSE REGULATORs* (*ARRs*) that are repressors of cytokinin signaling (To et al. [2004]). Similarly, the WUSCHEL-related Homeobox transcription factor WOX11 controls the expression of different *ARRs* and regulates both crown root initiation and development (Zhao et al. [2009]). *WOX11* expression is induced by auxin and cytokinin. This result shows that *CRL5* and *WOX11* are involved in the crossway of the antagonistic auxin and cytokinin pathways in the regulation of crown-root initiation.

The *HEME OXYGENASE 1* (*HOX1*) gene controls the formation of lateral roots via the production of carbon monoxide. *HOX1* is regulated by auxin and stress related signals, such as jasmonate and nitric oxide (Chen et al. [2012]; Hsu et al. [2012]), which suggests that this pathway can contribute to the modulation of root architecture in response to stress.

Genes Involved in Root Meristem Formation

After the initiation step, cells divide and become organized to shape a root meristem. In root meristems, the quiescent center (QC), which is composed of a small number of mitotically inactive cells, is indispensable for the maintenance of the undifferentiated status of the surrounding stem cells from which root tissues are produced (van den Berg et al. [1997]). The maintenance by auxin of the QC in rice is mediated by OsIAA23 (Jun et al. [2011]). The *QUIESCENT CENTER HOMEOBOX*(*QHB*) gene is an ortholog to the *WUSCHEL*-related *WOX5*, a QC-specific gene in *Arabidopsis* that contributes to QC and root stem cell maintenance (Kamiya et al. [2003b]; Breuninger et al.[2008]; Ditengou et al. [2008]). *QHB* is expressed early during root formation in the QC, and its expression is dependent on *CRL1/ARL1* during crown root formation (Inukai et al. [2005]; Ni et al.[2014]; Kamiya et al. [2003b]; Liu et al. [2005]). The CLAVATA3 (CLV3)/ENDOSPERM SURROUNDING REGION (ESR)-related (CLE) FON2-LIKE CLE PROTEIN1 (FCP1) and FCP2 proteins inhibit the expression of *QHB* in the QC (Chu et al. [2013]; Suzaki et al. [2008]). This relationship suggests that a CLAVATA/WOX signaling module regulates the activity of the root apical meristem in rice. Such a module has already been described in *Arabidopsis* for the regulation of the shoot apical meristem activity. In this context, the transduction of the CLV3 peptide signal is ensured by LEUCINE RICH REPEAT RECEPTOR-LIKE KINASE (LRR RLK) CLV1 (Muller et al. [2006]). In rice lines overexpressing *OsRPK1*, which encodes an LRR RLK, the expression levels of different *OsPIN*genes are down regulated, polar auxin transport is perturbed and the length of the seminal root is reduced, as is the number of crown roots and the density of lateral roots (Zou et al. [2014]). Nevertheless, it is undetermined at which step of root development (i.e., initiation, meristem maintenance or other

Pi: Inorganic phosphate
PIN: Pinformed auxin efflux carrier proteins
RPK1: Receptor protein kinase 1
PSTOL1: Phosphorus-Starvation tolerance 1
QHB: Quiescent center homeobox
RSS: Rice salt sensitive
SCR: Scarecrow
SHR: Shortroot
SPR1: Short postembryonic roots
TIR1: Transport inhibitor response 1
WOX: Wushel-related homeobox

Competing interests

The authors declare that they have no competing interests.

Authors' contributions

All authors contributed to collect information and write the manuscript.

ACKNOWLEDGEMENTS

NTPP is supported by a Global Rice International Science Partnership PhD fellowship, MG by a fellowship of IRD and the Ministère des Affaires Etrangères Français. The Vietnamese Ministry of Agriculture and Rural Development supports our research on root development in Vietnam.

REFERENCES

1. Abe M, Kuroshita H, Umeda M, Itoh J, Nagato Y (2008) The rice flattened shoot meristem, encoding CAF-1 p150 subunit, is required for meristem maintenance by regulating the cell-cycle period. Dev Biol 319:384-393

2. Bailey-Serres J, Fukao T, Ronald P, Ismail A, Heuer S, Mackill D (2010) Submergence tolerant rice: SUB1's journey from landrace to modern cultivar. Rice 3:138-147
3. Bian H, Xie Y, Guo F, Han N, Ma S, Zeng Z, Wang J, Yang Y, Zhu M (2012) Distinctive expression patterns and roles of the *miRNA393/TIR1* homolog module in regulating flag leaf inclination and primary and crown root growth in rice (*Oryza sativa*). New Phytol 196:149-161
4. Breuninger H, Rikirsch E, Hermann M, Ueda M, Laux T (2008) Differential expression of *WOX* genes mediates apical-basal axis formation in the Arabidopsis embryo. Dev Cell 14:867-876
5. Champoux M, Wang G, Sarkarung S, Mackill DJ, O'Toole JC, Huang N, McCouch SR (1995) Locating genes associated with root morphology and drought avoidance in rice via linkage to molecular marker. Theor Appl Genet 90:969-981
6. Chen JS, Lin SC, Chen CY, Hsieh YT, Pai PH, Chen LK, Lee S (2014) Development of a microarray for two rice subspecies: characterization and validation of gene expression in rice tissues. BMC Res Notes 7:15
7. Chen X, Shi J, Hao X, Liu H, Wu Y, Wu Z, Chen M, Wu P, Mao C (2013) OsORC3 is required for lateral root development in rice. Plant J 74:339-350
8. Chen YH, Chao YY, Hsu YY, Hong CY, Kao CH (2012) Heme oxygenase is involved in nitric oxide- and auxin-induced lateral root formation in rice. Plant Cell Rep 31:1085-1091
9. Cheng Y, Dai X, Zhao Y (2004) AtCAND1, a HEAT-repeat protein that participates in auxin signaling in Arabidopsis. Plant Physiol 135:1020-1026
10. Cho SH, Yoo SC, Zhang H, Pandeya D, Koh HJ, Hwang JY, Kim GT, Paek NC (2013) The rice*narrow leaf2* and *narrow leaf3* loci encode WUSCHEL-related homeobox 3A (OsWOX3A) and function in leaf, spikelet, tiller and lateral root development. New Phytol 198:1071-1084
11. Chu H, Liang W, Li J, Hong F, Wu Y, Wang L, Wang J, Wu P, Liu C, Zhang Q, Xu J, Zhang D (2013) A CLE-WOX signalling module regulates root meristem maintenance and vascular tissue development in rice. J Exp Bot 64:5359-5369

12. Clark RT, Famoso AN, Zhao K, Shaff JE, Craft EJ, Bustamante CD, McCouch SR, Aneshansley DJ, Kochian LV (2013) High-throughput two-dimensional root system phenotyping platform facilitates genetic analysis of root growth and development. Plant Cell Environ 36:454-466

13. Coudert Y, Bes M, Le Thi VA, Pre M, Guiderdoni E, Gantet P (2011) Transcript profiling of *crown root less 1* stem base reveals new elements associated with crown root development in rice. BMC Genomics 12:e387

14. Coudert Y, Dievart A, Droc G, Gantet P (2013) ASL/LBD phylogeny suggests that genetic mechanisms of root initiation downstream of auxin are distinct in lycophytes and euphyllophytes. Mol Biol Evol 30:569-572

15. Coudert Y, Le Thi VA, Gantet P (2013b) Rice: a model plant to decipher the hidden origin of adventitious roots. In: Eshel A, Beeckman T (eds) Plant roots: the hidden half, 4th edn. CCR press, Taylor and Francis group, LLC, Boca Raton, USA. pp 157-166

16. Coudert Y, Perin C, Courtois B, Khong NG, Gantet P (2010) Genetic control of root development in rice, the model cereal. Trends Plant Sci 15:219-226

17. Courtois B, Ahmadi N, Khowaja F, Price AH, Rami J-F, Frouin J, Hamelin C, Ruiz M (2009) Rice root genetic architecture: meta-analysis from a drought QTL database. Rice 2:115-128

18. Courtois B, Audebert A, Dardou A, Roques S, Ghneim-Herrera T, Droc G, Frouin J, Rouan L, Goze E, Kilian A, Ahmadi N, Dingkuhn M (2013) Genome-wide association mapping of root traits in a japonica rice panel. PLoS One 8:e78037

19. Cui H, Levesque MP, Vernoux T, Jung JW, Paquette AJ, Gallagher KL, Wang JY, Blilou I, Scheres B, Benfey PN (2007) An evolutionarily conserved mechanism delimiting SHR movement defines a single layer of endodermis in plants. Science 316:421-425

20. Den Herder G, Van Isterdael G, Beeckman T, De Smet I (2010) The roots of a new green revolution. Trends Plant Sci 15:600-607

21. Di Laurenzio L, Wysocka-Diller J, Malamy JE, Pysh L, Helariutta Y, Freshour G, Hahn MG, Feldmann KA, Benfey PN (1996) The *SCARECROW* gene regulates an asymmetric cell division

that is essential for generating the radial organization of the Arabidopsis root. Cell 86:423-433

22. Ditengou FA, Teale WD, Kochersperger P, Flittner KA, Kneuper I, van der Graaff E, Nziengui H, Pinosa F, Li X, Nitschke R, Laux T, Palme K (2008) Mechanical induction of lateral root initiation in*Arabidopsis thaliana*. Proc Natl Acad Sci U S A 105:18818-18823
23. Fukai S, Cooper M (1995) Development of drought resistant cultivars using physio-morphological traits in rice. Field Crops Res 40:67-87
24. Gamuyao R, Chin JH, Pariasca-Tanaka J, Pesaresi P, Catausan S, Dalid C, Slamet-Loedin I, Tecson-Mendoza EM, Wissuwa M, Heuer S (2012) The protein kinase Pstol1 from traditional rice confers tolerance of phosphorus deficiency. Nature 488:535-539
25. Gowda VRP, Henry A, Yamauchi A, Shashidhar HE, Serraj R (2011) Root biology and genetic improvement for drought avoidance in rice. Field Crop Res 122:1-13
26. Guiderdoni E, Gantet P (2012) Ac-Ds solutions for rice insertion mutagenesis. Methods Mol Biol 859:177-187
27. Guilfoyle TJ, Hagen G (2007) Auxin response factors. Curr Opin Plant Biol 10(5):453-460
28. Hirochika H, Guiderdoni E, An G, Hsing YI, Eun MY, Han CD, Upadhyaya N, Ramachandran S, Zhang Q, Pereira A, Sundaresan V, Leung H (2004) Rice mutant resources for gene discovery. Plant Mol Biol 54:325-334
29. Hochholdinger F, Tuberosa R (2009) Genetic and genomic dissection of maize root development and architecture. Curr Opin Plant Biol 12:172-177
30. Hsu YY, Chao YY, Kao CH (2012) Methyl jasmonate-induced lateral root formation in rice: The role of heme oxygenase and calcium. Journal of plant physiol 170:63-69
31. Huang CF, Yamaji N, Nishimura M, Tajima S, Ma JF (2009) A rice mutant sensitive to Al toxicity is defective in the specification of root outer cell layers. Plant Cell Physiol 50:976-985
32. Huang CF, Yamaji N, Ono K, Ma JF (2012) A leucine-rich repeat receptor-like kinase gene is involved in the specification of outer cell layers in rice roots. Plant J 69:565-576

33. Inukai Y, Sakamoto T, Ueguchi-Tanaka M, Shibata Y, Gomi K, Umemura I, Hasegawa Y, Ashikari M, Kitano H, Matsuoka M (2005) *Crown rootless1*, which is essential for crown root formation in rice, is a target of an AUXIN RESPONSE FACTOR in auxin signaling. Plant Cell 17:1387-1396

34. Itoh J, Nonomura K, Ikeda K, Yamaki S, Inukai Y, Yamagishi H, Kitano H, Nagato Y (2005) Rice plant development: from zygote to spikelet. Plant Cell Physiol 46:23-47

35. Itoh T, Tanaka T, Barrero RA, Yamasaki C, Fujii Y, Hilton PB, Antonio BA, Aono H, Apweiler R, Bruskiewich R, Bureau T, Burr F, Costa de Oliveira A, Fuks G, Habara T, Haberer G, Han B, Harada E, Hiraki AT, Hirochika H, Hoen D, Hokari H, Hosokawa S, Hsing YI, Ikawa H, Ikeo K, Imanishi T, Ito Y, Jaiswal P, Kanno M, Kawahara Y, Kawamura T, Kawashima H, Khurana JP, Kikuchi S, Komatsu S, Koyanagi KO, Kubooka H, Lieberherr D, Lin YC, Lonsdale D, Matsumoto T, Matsuya A, McCombie WR, Messing J, Miyao A, Mulder N, Nagamura Y, Nam J, Namiki N, Numa H, Nurimoto S, O'Donovan C, Ohyanagi H, Okido T, Oota S, Osato N, Palmer LE, Quetier F, Raghuvanshi S, Saichi N, Sakai H, Sakai Y, Sakata K, Sakurai T, Sato F, Sato Y, Schoof H, Seki M, Shibata M, Shimizu Y, Shinozaki K, Shinso Y, Singh NK, Smith-White B, Takeda J, Tanino M, Tatusova T, Thongjuea S, Todokoro F, Tsugane M, Tyagi AK, Vanavichit A, Wang A, Wing RA, Yamaguchi K, Yamamoto M, Yamamoto N, Yu Y, Zhang H, Zhao Q, Higo K, Burr B, Gojobori T, Sasaki T (2007) Curated genome annotation of *Oryza sativa* ssp. japonica and comparative genome analysis with*Arabidopsis thaliana*. Genome Res 17:175-183

36. Jeong JS, Kim YS, Baek KH, Jung H, Ha SH, Do Choi Y, Kim M, Reuzeau C, Kim JK (2010) Root-specific expression of *OsNAC10* improves drought tolerance and grain yield in rice under field drought conditions. Plant Physiol 153:185-197

37. Jeong JS, Kim YS, Redillas MC, Jang G, Jung H, Bang SW, Choi YD, Ha SH, Reuzeau C, Kim JK (2013) *OsNAC5* overexpression enlarges root diameter in rice plants leading to enhanced drought tolerance and increased grain yield in the field. Plant Biotechnol J 11(1):101-114

38. Jia L, Wu Z, Hao X, Carrie C, Zheng L, Whelan J, Wu Y, Wang S, Wu P, Mao C (2011) Identification of a novel mitochondrial

protein, short postembryonic roots 1 (SPR1), involved in root development and iron homeostasis in *Oryza sativa*. New Phytol 189:843-855

39. Jun N, Gaohang W, Zhenxing Z, Huanhuan Z, Yunrong W, Ping W (2011) OsIAA23-mediated auxin signaling defines postembryonic maintenance of QC in rice. Plant J 68:433-442
40. Kamiya N, Itoh J, Morikami A, Nagato Y, Matsuoka M (2003) The *SCARECROW* gene's role in asymmetric cell divisions in rice plants. Plant J 36:45-54
41. Kamiya N, Nagasaki H, Morikami A, Sato Y, Matsuoka M (2003) Isolation and characterization of a rice WUSCHEL-type homeobox gene that is specifically expressed in the central cells of a quiescent center in the root apical meristem. Plant J 35:429-441
42. Kamoshita A, Chandra Babu R, Boopathi MN, Fukai S (2008) Phenotypic and genotypic analysis of drought-resistance traits for development of rice cultivars adapted to rainfed environments. Field Crop Research 109:1-23
43. Kamoshita A, Zhang J, Siopongco J, Sarkarung S, Nguyen HT, Wade LJ (2002) Effects of phenotyping environment on identification of quantitative trait loci for rice root morphology under anaerobic conditions. Crop Sci 42:255-265
44. Kang B, Zhang Z, Wang L, Zheng L, Mao W, Li M, Wu Y, Wu P, Mo X (2013) OsCYP2, a chaperone involved in degradation of auxin-responsive proteins, plays crucial roles in rice lateral root initiation. Plant J 74:86-97
45. Khowaja FS, Norton GJ, Courtois B, Price AH (2009) Improved resolution in the position of drought-related QTLs in a single mapping population of rice by meta-analysis. BMC Genomics 10:e276
46. Kitomi Y, Inahashi H, Takehisa H, Sato Y, Inukai Y (2012) OsIAA13-mediated auxin signaling is involved in lateral root initiation in rice. Plant Sci 190:116-122
47. Kitomi Y, Ito H, Hobo T, Aya K, Kitano H, Inukai Y (2011) The auxin responsive AP2/ERF transcription factor CROWN ROOTLESS5 is involved in crown root initiation in rice through the induction of

OsRR1, a type-A response regulator of cytokinin signaling. Plant J 67:472-484

48. Kitomi Y, Ogawa A, Kitano H, Inukai Y (2008) *CRL4* regulates crown root formation through auxin transport in rice. Plant Root 2:19-28

49. Khowaja FS, Norton GJ, Courtois B, Price AH (2009) Improved resolution in the position of drought-related QTLs in a single mapping population of rice by meta-analysis. BMC Genomics 10(276):1-14

50. Krishnan A, Guiderdoni E, An G, Hsing YI, Han CD, Lee MC, Yu SM, Upadhyaya N, Ramachandran S, Zhang Q, Sundaresan V, Hirochika H, Leung H, Pereira A (2009) Mutant resources in rice for functional genomics of the grasses. Plant Physiol 149:165-170

51. Lafitte H, Champoux M, McLaren G, O'Toole J (2001) Rice root morphological traits are related to isozyme group and adaptation. Field Crop Research 71:57-70

52. Lavenus J, Goh T, Roberts I, Guyomarc'h S, Lucas M, De Smet I, Fukaki H, Beeckman T, Bennett M, Laplaze L (2013) Lateral root development in Arabidopsis: fifty shades of auxin. Trends Plant Sci 18:450-458

53. Liu H, Wang S, Yu X, Yu J, He X, Zhang S, Shou H, Wu P (2005) ARL1, a LOB-domain protein required for adventitious root formation in rice. Plant J 43:47-56 PubMed Abstract |Publisher Full Text

54. Liu S, Wang J, Wang L, Wang X, Xue Y, Wu P, Shou H (2009) Adventitious root formation in rice requires OsGNOM1 and is mediated by the OsPINs family. Cell Res 19:1110-1119

55. Lorbiecke R, Sauter M (1999) Adventitious root growth and cell-cycle induction in deepwater rice. Plant Physiol 119:21-30

56. Lorieux M, Blein M, Lozano J, Bouniol M, Droc G, Dievart A, Perin C, Mieulet D, Lanau N, Bes M, Rouviere C, Gay C, Piffanelli P, Larmande P, Michel C, Barnola I, Biderre-Petit C, Sallaud C, Perez P, Bourgis F, Ghesquiere A, Gantet P, Tohme J, Morel JB, Guiderdoni E (2012) In-depth molecular and phenotypic characterization in a rice insertion line library facilitates gene identification through reverse and forward genetics approaches. Plant Biotechnol J 10:555-568

57. Ma N, Wang Y, Qiu S, Kang Z, Che S, Wang G, Huang J (2013) Overexpression of *OsEXPA8*, a root-specific gene, improves rice growth and root system architecture by facilitating cell extension. PLoS One 8:e75997

58. MacMillan K, Emrich K, Piepho HP, Mullins CE, Price AH (2006) Assessing the importance of genotype x environment interaction for root traits in rice using a mapping population II: conventional QTL analysis. Theor Appl Genet 113:953-964

59. Mergemann H, Sauter M (2000) Ethylene induces epidermal cell death at the site of adventitious root emergence in rice. Plant Physiol 124:609-614

60. Mockaitis K, Estelle M (2008) Auxin receptors and plant development: a new signaling paradigm. Annu Rev Cell Dev Biol 24:55-80

61. Muller R, Borghi L, Kwiatkowska D, Laufs P, Simon R (2006) Dynamic and compensatory responses of Arabidopsis shoot and floral meristems to CLV3 signaling. Plant Cell 18:1188-1198

62. Ni J, Shen Y, Zhang Y, Wu P (2014) Definition and stabilisation of the quiescent center in rice roots. Plant Biol, in press

63. O'Toole J, Bland W (1987) Genotypic variation in crop plant root systems. Adv Agron 41:91-145

64. Ogawa D, Abe K, Miyao A, Kojima M, Sakakibara H, Mizutani M, Morita H, Toda Y, Hobo T, Sato Y, Hattori T, Hirochika H, Takeda S (2011) RSS1 regulates the cell cycle and maintains meristematic activity under stress conditions in rice. Nat Commun 2:278

65. Orman-Ligeza B, Parizot B, Gantet P, Beeckman T, Bennett MJ, Draye X (2013) Post-embryonic root organogenesis in cereals: branching out from model plants. Trends Plant Sci 18:459-467

66. Peret B, De Rybel B, Casimiro I, Benkova E, Swarup R, Laplaze L, Beeckman T, Bennett MJ (2009) Arabidopsis lateral root development: an emerging story. Trends Plant Sci 14:399-408

67. Puig J, Pauluzzi G, Guiderdoni E, Gantet P (2012) Regulation of shoot and root development trought mutual signaling. Mol Plant 5:574-583

68. Qin C, Li Y, Gan J, Wang W, Zhang H, Liu Y, Wu P (2013) OsDGL1, a homolog of an oligosaccharyltransferase complex subunit, is

involved in N-glycosylation and root development in rice. Plant Cell Physiol 54:129-137

69. Rani Debi B, Taketa S, Ichii M (2005) Cytokinin inhibits lateral root initiation but stimulates lateral root elongation in rice (*Oryza sativa*). J Plant Physiol 162:507-515

70. Rebouillat A, Dievart A, Verdeil JL, Escoute J, Giese G, Breitler JC, Gantet P, Espeout S, Guiderdoni E, Périn C (2009) Molecular genetic of rice root development. Rice 2:15-34

71. Redillas MC, Jeong JS, Kim YS, Jung H, Bang SW, Choi YD, Ha SH, Reuzeau C, Kim JK (2012) The overexpression of *OsNAC9* alters the root architecture of rice plants enhancing drought resistance and grain yield under field conditions. Plant Biotechnol J 10:792-805

72. Sabatini S, Heidstra R, Wildwater M, Scheres B (2003) SCARECROW is involved in positioning the stem cell niche in the Arabidopsis root meristem. Genes Dev 17:354-358

73. Sato Y, Takehisa H, Kamatsuki K, Minami H, Namiki N, Ikawa H, Ohyanagi H, Sugimoto K, Antonio BA, Nagamura Y (2013) RiceXPro version 3.0: expanding the informatics resource for rice transcriptome. Nucleic Acids Res 41(Database issue):D1206-D1213

74. Schatz MC, Maron LG, Stein JC, Hernandez Wentz A, Gurtowski J, Biggers E, Lee H, Kramer M, Antoniou E, Ghiban E, Wright MH, Chia JM, Ware DM, McCouch SR, McCombie WR (2014) New whole genome de novo assemblies of three divergent strains of rice documents novel gene space of aus and indica. bioRxiv, in press

75. Shen L, Courtois B, McNally K, Robin S, Li Z (2001) Evaluation of near-isogenic lines of rice introgressed with QTLs for root depth through marker-aided selection. Theor Appl Genet 103:75-83

76. Shin JH, Jeong DH, Park MC, An G (2005) Characterization and transcriptional expression of the alpha-expansin gene family in rice. Mol Cells 20:210-218

77. Steele K, Price A, Shashidhar H, Witcombe J (2006) Marker-assisted selection to introgress rice QTLs controlling root traits into an Indian upland rice variety. Theor Appl Genet 112:208-221

78. Steele KA, Price AH, Witcombe JR, Shrestha R, Singh BN, Gibbons JM, Virk DS (2013) QTLs associated with root traits increase yield in upland rice when transferred through marker-assisted selection. Theor Appl Genet 126:101-108

79. Steffens B, Wang J, Sauter M (2006) Interactions between ethylene, gibberellin and abscisic acid regulate emergence and growth rate of adventitious roots in deepwater rice. Planta 223:604-612

80. Steinmann T, Geldner N, Grebe M, Mangold S, Jackson CL, Paris S, Galweiler L, Palme K, Jurgens G (1999) Coordinated polar localization of auxin efflux carrier PIN1 by GNOM ARF GEF. Science 286:316-318

81. Suji KK, Prince KSJ, Mankhar PS, Kanagaraj P, Poornima R, Amutha K, Kavitha S, Biji KR, Gomez SM, Chandra Babu R (2012) Evaluation of rice near isogenic lines with root QTLs for plant production and root traits in rainfed target populations of environment. Field Crop Res 137:89-96

82. Suzaki T, Yoshida A, Hirano HY (2008) Functional diversification of CLAVATA3-related CLE proteins in meristem maintenance in rice. Plant Cell 20:2049-2058

83. Sweeney M, McCouch S (2007) The complex history of the domestication of rice. Ann Bot 100:951-957

84. Takehisa H, Sato Y, Igarashi M, Abiko T, Antonio BA, Kamatsuki K, Minami H, Namiki N, Inukai Y, Nakazono M, Nagamura Y (2012) Genome-wide transcriptome dissection of the rice root system: implications for developmental and physiological functions. Plant J 69:126-140

85. To JP, Haberer G, Ferreira FJ, Deruere J, Mason MG, Schaller GE, Alonso JM, Ecker JR, Kieber JJ (2004) Type-A Arabidopsis response regulators are partially redundant negative regulators of cytokinin signaling. Plant Cell 16:658-671

86. Toda Y, Tanaka M, Ogawa D, Kurata K, Kurotani K, Habu Y, Ando T, Sugimoto K, Mitsuda N, Katoh E, Abe K, Miyao A, Hirochika H, Hattori T, Takeda S (2013) RICE SALT SENSITIVE3 forms a ternary complex with JAZ and class-C bHLH factors and regulates jasmonate-induced gene expression and root cell elongation. Plant Cell 25:1709-1725

87. Topp CN, Iyer-Pascuzzi AS, Anderson JT, Lee CR, Zurek PR, Symonova O, Zheng Y, Bucksch A, Mileyko Y, Galkovskyi T, Moore BT, Harer J, Edelsbrunner H, Mitchell-Olds T, Weitz JS, Benfey PN (2013) 3D phenotyping and quantitative trait locus mapping identify core regions of the rice genome controlling root architecture. Proc Natl Acad Sci U S A 110:1695-1704

88. Uga Y, Sugimoto K, Ogawa S, Rane J, Ishitani M, Hara N, Kitomi Y, Inukai Y, Ono K, Kanno N, Inoue H, Takehisa H, Motoyama R, Nagamura Y, Wu J, Matsumoto T, Takai T, Okuno K, Yano M (2013) Control of root system architecture by DEEPER ROOTING 1 increases rice yield under drought conditions. Nat Genet 45:1097-1102

89. van den Berg C, Willemsen V, Hendriks G, Weisbeek P, Scheres B (1997) Short-range control of cell differentiation in the Arabidopsis root meristem. Nature 390:287-289

90. Vanneste S, Friml J (2009) Auxin: a trigger for change in plant development. Cell 136:1005-1016PubMed Abstract | Publisher Full Text

91. Wang XF, He FF, Ma XX, Mao CZ, Hodgman C, Lu CG, Wu P (2010) OsCAND1 is required for crown root emergence in rice. Mol Plant 4:289-299

92. Wei FJ, Droc G, Guiderdoni E, Hsing YI (2013) International consortium of rice mutagenesis: resources and beyond. Rice: 6:39

93. Xia K, Wang R, Ou X, Fang Z, Tian C, Duan J, Wang Y, Zhang M (2012) *OsTIR1* and *OsAFB2*downregulation via *OsmiR393* overexpression leads to more tillers, early flowering and less tolerance to salt and drought in rice. PLoS One 7:e30039

94. Xiong J, Tao L, Zhu C (2009) Does nitric oxide play a pivotal role downstream of auxin in promoting crown root primordia initiation in monocots? Plant Signal Behav 4:999-1001

95. Xu M, Zhu L, Shou H, Wu P (2005) A PIN1 family gene, OsPIN1, involved in auxin-dependent adventitious root emergence and tillering in rice. Plant Cell Physiol 46:1674-1681

96. Yoo SC, Cho SH, Paek NC (2013) Rice WUSCHEL-related homeobox 3A (OsWOX3A) modulates auxin-transport gene

flowering have been identified, the process by which rice eventually flowers under non-permissive LD conditions is not well understood.

Results

Overexpression of miR172 reduced flowering time significantly, suggesting its role as an inducer. Levels of miR172 increased as plants aged, further supporting our findings. Transcripts of SNB andOsIDS1, two members of the AP2 family that have the miR172 target site, were reduced in older plants as the level of miR172 rose. Overexpression of those AP2 genes delayed flowering; overexpression of miR172-resistant forms of SNB or OsIDS1 further delayed this process. This demonstrated that the AP2 genes function downstream of miR172. Two florigen genes -- Hd3a andRFT1 -- and their immediate upstream regulator Ehd1 were suppressed in the AP2 overexpression plants. This suggested that the AP2 genes are upstream repressors of Ehd1. In phytochrome mutants, miR172d levels were increased whereas those of SNB and OsIDS1 were decreased. Thus, it appears that phytochromes inhibit miR172d, an AP2 suppresser.

Conclusions

We revealed that *miR172d* developmentally induced flowering via repressing *OsIDS1* and *SNB*, which suppressed *Ehd1*. We also showed that phytochromes negatively regulated *miR172*.

BACKGROUND

Rice flowers earlier under short day (SD) conditions than under long days (LD). The photoperiodic flowering pathway of rice is controlled by *OsGIGANTEA* (*OsGI*) (Yano et al. [2000]; Hayama et al.[2002], [2003]; Kojima et al. [2002]; Doi et al. [2004]). This gene regulates flowering time by promoting *Heading date 1* (*Hd1*), an ortholog of *Arabidopsis CONSTANS* (*CO*) (Yano et al. [2000]; Hayama et al. [2002], [2003]). *Hd1* enhances *Early heading date 1* (*Ehd1*) expression under SD but inhibits its expression under LD (Wei et al. [2010]; Ishikawa et al. [2011]). *Ehd1* is an immediate upstream positive regulator of

Heading date 3a (*Hd3a*) and *Rice Flowering Locus T 1* (*RFT1*), which encode florigens that promote flowering (Doi et al. [2004]; Tamaki et al. [2007]; Komiya et al.[2008], [2009]).

Because *Ehd1* is modulated by several regulators, it acts as a mediator of various floral signals. For example, *Early heading date 2* (*Ehd2*)/*OsINDETERMINATE 1* (*OsId1*)/*Rice INDETERMINATE 1*(*RID1*) is a constitutive activator of *Ehd1* (Matsubara et al. [2008]; Park et al. [2008]; Wu et al.[2008]). Loss-of-function mutants in the gene do not flower under either SD or LD. *OsMADS51*also acts as a positive regulator of *Ehd1*, specifically under SD (Kim et al. [2007]). One major repressor of *Ehd1* is *Grain yield and heading date 7* (*Ghd7*), which functions preferentially under LD (Xue et al. [2008]). Most early-flowering rice cultivars show a disruption in *Ghd7* expression (Xue et al. [2008]). A chromatin remodeling factor, *OsTrithorax 1* (*OsTrx1*), suppresses *Ghd7* by binding to*Early heading date 3* (*Ehd3*) (Matsubara et al. [2011]; Choi et al. [2014]). In addition, *OsLFL1*inhibits *Ehd1* when over-expressed (Peng et al. [2007], [2008]). However, expression of the former is suppressed by another chromatin remodeling factor, *OsVIN3-like 2* (*OsVIL2*), when binding to the PRC2 complex (Yang et al. [2013]). *OsMADS50* induces *Ehd1* by blocking *OsLFL1* and *Ghd7* (Lee et al. [2004]; Ryu et al. [2009]; Choi et al. [2014]). A third gene, *OsCOL4*, constitutively inhibits flowering time by suppressing *Ehd1* (Lee et al. [2010]). Overexpression of the former delays flowering whereas knockout mutations cause early flowering. Finally, *DTH8* and *Hd16* preferentially suppress flowering time under LD by inhibiting *Ehd1* (Xue et al. [2008]; Wei et al. [2010]; Hori et al.[2013]).

Micro RNAs inhibit expression of target genes by cleaving mRNA or translational suppression (Jones-Rhoades et al. [2006];Voinnet [2009]). *miR172* and *miR156* are involved in phase transition (Aukerman and Sakai [2003]; Lauter et al. [2005]; Wu and Poethig [2006]; Poethig[2009]). *miR156* plays roles in early vegetative stages, while *miR172* functions later stages of develop (Aukerman and Sakai [2003]; Lauter et al. [2005]; Wu and Poethig [2006]; Chuck et al.[2007]; Poethig [2009]). In *Arabidopsis*, *miR156* targets 10 members (*SPL2*, *SPL3*, *SPL4*, *SPL5*,*SPL6*, *SPL9*, *SPL10*, *SPL11*, *SPL13*, and *SPL15*) of *SQUAMOSA PROMOTER BINDING PROTEIN-LIKE*(*SPL*) family. *SPL9* prompts *miR172* expression and the other *SPL* genes redundantly function in regulating *miR172*.The role of *miR172* in controlling flowering time has been reported for*Arabidopsis*, maize, barley, and soybean (Aukerman and Sakai [2003]; Chen [2004]; Lauter et al.[2005]; Jung et al. [2007]; Mathieu et al. [2009]; Nair et al. [2010]; Yoshikawa et al. [2013]). In*Arabidopsis* and maize, its temporal expression increases gradually as plants age (Aukerman and Sakai [2003]; Lauter et al. [2005]). In rice, *miR172* is most highly expressed during later vegetative stages and in developing panicles (Zhu et al. [2009]; Lee and An [2012]).

AP2 family genes are involved in various processes, including floral organ identity, shattering, and flowering time (Aukerman and Sakai [2003]; Chen [2004]; Lauter et al. [2005]; Lee et al. [2007]; Jung et al. [2007]; Chuck et al. [2008]; Mathieu et al. [2009]; Zhu et al. [2009]; Lee and An[2012]; Zhou et al. [2012]; Yoshikawa et al. [2013]). Six *Arabidopsis* genes in this family --*APETALA 2* (*AP2*), *TARGET OF EAT 1*(*TOE1*), *TOE2*, *TOE3*, *SCHLAFMUTZE* (*SMZ*), and*SCHNARCHZAPFEN* (*SNZ*) -- delay flowering in an age-dependent manner (Park et al. [2002]; Aukerman and Sakai [2003]; Schmid et al. [2003]; Chen [2004]; Jung et al. [2007]; Mathieu et al.[2009]), and are suppressed by *miR172* (Schmid et al. [2003]; Kasschau et al. [2003]; Chen[2004]; Schwab et al. [2005]; Jung et al. [2007]; Mathieu et al. [2009]). In maize, enhancement of*GLOSSY15* (*GL15*), an *AP2* member, delays phase transition from the vegetative to the reproductive stage (Lauter et al. [2005], Zhu and Helliwell [2011]). Its temporal expression gradually decreases as plants mature, and this gene is also down-regulated by *miR172* (Lauter et al. [2005]). In rice, five AP2-like genes (*SNB*, *OsIDS1*, *SHAT1*, *Os05g03040*, and *Os06g43220*) contain the *miR172*target sites (Sunkar et al. [2005]; Zhu et al. [2009]). While *SNB* and *OsIDS1* control floral organ identity and spikelet development, *SHAT1* is involved in seed shattering (Sunkar et al. [2005]; Lee et al. [2007]; Zhu et al. [2009]; Lee and An [2012]; Zhou et al. [2012]).

Although an antagonistic role for *miR172* and AP2 genes in floral transition has been described in*Arabidopsis* and maize, their functions in rice have not been reported. Here, we demonstrated in rice that *miR172* induces flowering time by suppressing two *AP2* family members -- *SNB* and *OsIDS1* – that are negative regulators of *Ehd1*.

RESULTS

Temporal Expression Patterns of miR172s and AP2 genes are antagonistic

To Examine Whether MiR172 Levels Change as Plants Develop, We Monitored the Temporal expressions of *miR172a* and *miR172d* in leaf blades. The other two *miR172* members in rice --*miR172b* and *miR172c* – were not measured because they are expressed mainly in panicles and roots, respectively, and are not likely involved in controlling flowering time (Jeong et al. [2011]). Levels of pri-*miR172a* and pri-*miR172d* increased gradually as plants aged (Figure 1A and B). To verify this, we conducted northern blot analysis of mature *miR172*. Because the sequences of mature *miR172a* and *miR172d* are identical, we were only able to measure the total amount of both micro RNAs. This analysis confirmed that mature *miR172ad* levels were gradually increased (Figure 1C).

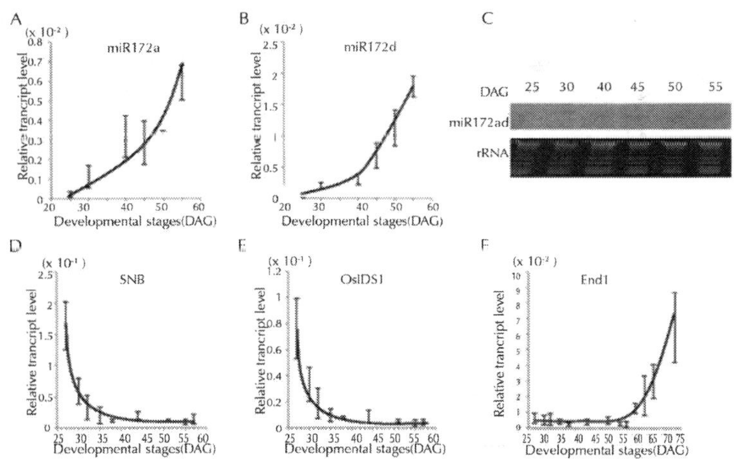

Figure 1: Temporal expression patterns of*miR172*and AP2 genes. (A, B) RT-PCR analysis of pri-*miR172a*(A) and pri-*miR172d*(B) in WT plants under LD. Y-axis, relative transcript level compared with *OsUbi1*. Error bars indicate standard deviations; *n* = 4 or more. Middle regions of fully emerged

uppermost leaves were sampled at 30, 40, 45, 50, 55, and 60 DAG. (C)Northern blot analysis of mature *miR172* (top). Ethidium bromide-stained ribosomal RNA served as loading control (bottom). Middle regions were used from leaves sampled at 25, 30, 40, 45, 50, and 55 DAG. (D, E) RT-PCR analysis of transcript levels of *SNB*(E) and*OsIDS1*(E) under LD. Middle regions were used from leaves sampled at 27, 30, 32, 35, 38, 44, 51, 55, and 57 DAG. (F) RT-PCR analysis of *Ehd1* transcripts under LD, using leaves sampled at 27, 30, 32, 35, 38, 44, 51, 55, 57, 60, 63, 66, and 73 DAG. Expression was monitored at ZT 18 h for *miR172* and at ZT 2 h for AP2 genes and *Ehd1*. Y-axis, relative transcript level compared with *OsUbi1*. Error bars indicate standard deviations; n = 4 or more.

The *miR172*s target several AP2 genes, with temporal expression of the latter type being slowly diminished in *Arabidopsis* and maize (Aukerman and Sakai [2003]; Lauter et al. [2005]; Jung et al.[2007]; Mathieu et al. [2009]; Zhu and Helliwell [2011]). In all, six AP2 genes have *miR172* target sites (Zhu and Helliwell [2011]). Other miRNAs and their targeted genes also show antagonistic expression patterns (Jeong and Green [2013]; Jung et al. [2013]). We examined the temporal expression of two rice AP2 genes in leaf blades: *SNB* and *OsIDS1*. As expected, their transcript levels were high at younger developmental stages but rapidly declined to a low level at 35 days after germination (DAG) (Figure 1D and E). *Ehd1* mRNA expression began to increase at 57 DAG (Figure 1F). Under SD, transcript levels of the AP2 genes dropped rapidly at 18 DAG, immediately before the level of *Ehd1* began to rise (Additional file 1: Figure S1). This suggested that the AP2 genes are likely targets of *miR172a* and *miR172d*.

Overexpression of *SNB* and *OsIDS1* Causes Late Flowering

To examine the functional roles of AP2 genes in flowering time, we generated transgenic rice plants that over-express *SNB* and *OsIDS1* (Additional file 1: Figure S2). Such overexpression (OX) of*miR172*-targeted AP2 causes late-flowering phenotypes in *Arabidopsis* (Jung et al. [2007]; Mathieu et al. [2009]). Similarly, we noted that *SNB* OX and *OsIDS1* OX plants flowered late when grown in the greenhouse (Figure 2A and B). Among our 11 *SNB* OX transgenics, flowering was delayed by 2 to 8 weeks for nine of them (Figure 2E). Flowering time for the *SNB* OX plants was correlated with the degree of *SNB* expression

(Additional file 1: Figure S2A, S2B, and S2C). Likewise, *OsIDS1* OX plants flowered 3 to 5 weeks later than usual (Figure 2F). Their heading date was correlated with amounts of *OsIDS1* transcript (Additional file 1: Figure S2D, S2E, and S2F). Three *SNB* OX plants (#3, 4, and 5) and three *OsIDS1* OX plants (#2, 3, and 6) were selected for further analysis. In studying photoperiod dependency of those genes, we found that flowering time for *SNB* OX #3 was delayed by about 12 d under SD and about 7 d under LD (Figure 2A, C, and D). Similarly, *OsIDS1*OX #2 flowered 3 weeks later than usual under SD and 2 weeks later under LD (Figure 2B, C, and D). These observations demonstrated that *SNB* and *OsIDS1* function as flowering repressors in rice.

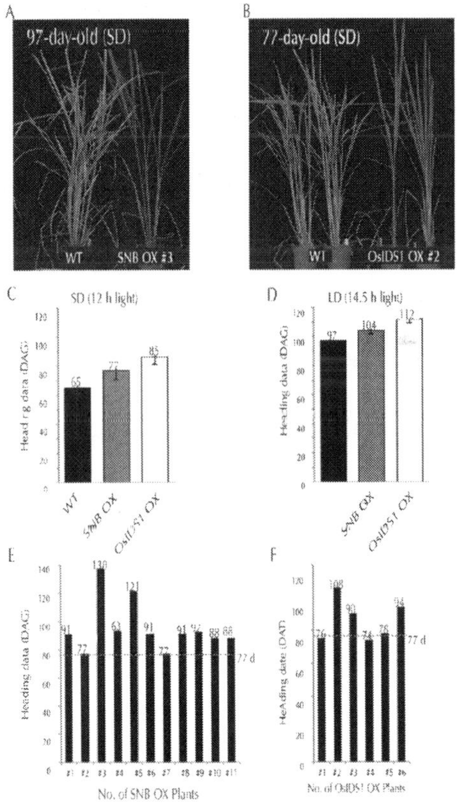

Figure 2 : Analysis of*SNB*and*OsIDS1*overexpression plants. (A)Phenotypes of *SNB* overexpression plant (*SNB* OX #3) at 97 DAG when grown under SD.

(B) Phenotypes of *OsIDS1* overexpression plants (*OsIDS1* OX #2) at 77 DAG when grown under SD. (C, D) Heading dates for *OsIDS1* and *SNB* OX plants under SD (C) or LD (D). Days-to-heading was scored when first panicle bolted. Error bars indicate standard deviations; n = 10-20 plants. (E, F) Flowering time for *SNB* (E) and*OsIDS1* (Ff) OX plants (T1 generation) in greenhouse. Days-to-heading was scored when first panicle bolted.

SNB and OsIDS1 repress the floral transition by inhibiting Ehd1

To elucidate the roles of *SNB* and *OsIDS1* in controlling flowering time, we measured transcript levels of the previously identified floral regulators in OX plants. Because the delayed-flowering phenotype was more severe in *OsIDS1* OX #2, we analyzed those plants. Increased expression of*OsIDS1* did not affect transcript levels of the AP2 genes (Figure 3B and C). Transcripts of *Hd3a* and*RFT1*, two florigens in rice, were significantly reduced in those plants (Figure 3D and E), as were transcripts of *Ehd1* (Figure 3F). However, levels of other flowering activators, i.e., *OsGI*, *Hd1*,*OsMADS51*, and *OsCO3*, were not altered in the OX plants (Figure 3G, H, I, and J). Overexpression of *OsIDS1* also had no influence on flowering time regulators such as *OsPhyB*, *OsMADS50*, *Ghd7*,*OsId1*, *OsCOL4*, *OsMADS56*, *Ehd3*, and *OsTrx1* (Additional file 1: Figure S3). In the *SNB* OX #3, transcript levels of *Ehd1*, *Hd3a*, and *RFT1* were also decreased, but those of *OsGI*, *Hd1*,*OsMADS51*, and *OsCOL4* were not changed (Additional file 1: Figure S4). These observations indicated that *SNB* and *OsIDS1* repress the floral transition by inhibiting the expression of *Ehd1*.

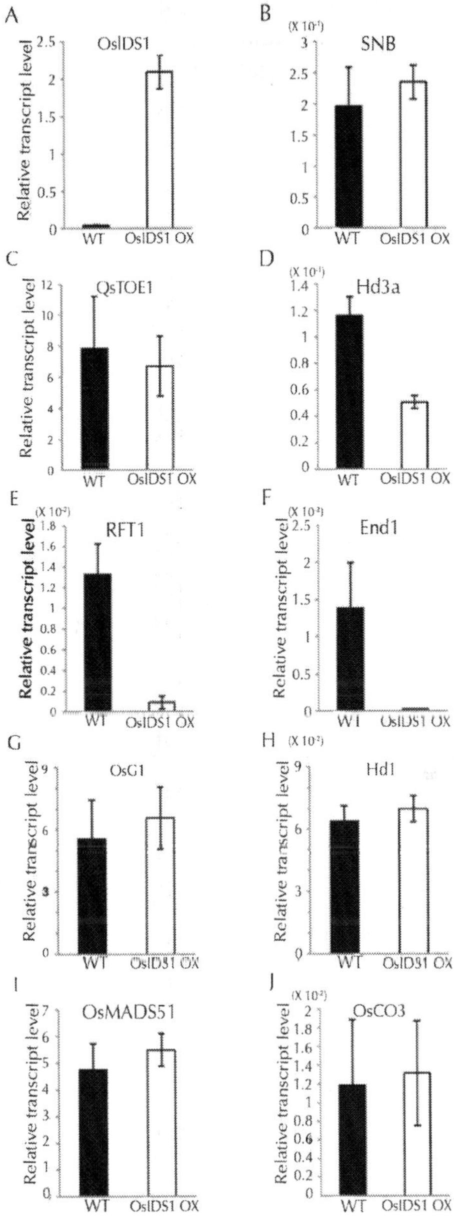

Figure 3: nTranscript levels of floral regulators in the*OsIDS1*OX.Transcript levels of *OsIDS1*(A), *SNB*(B), *OsTOE1*(C),*Hd3a*(D), *RFT1*(E),*Ehd1*(F), *OsGI*(G), *Hd1*(H), *OsMADS51*(I), and *OsCO3*(J) in WT (closed boxes) and *OsIDS1* OX

plants (open boxes) under SD. Middle regions of fully emerged uppermost leaves were sampled at 30 DAG. For *OsGI* and*Hd1*, monitoring occurred at ZT 10 h and ZT 14 h, respectively; for others, at ZT 2 h. Y-axis, relative transcript level compared with rice*OsUbi1*. Error bars indicate standard deviations; $n = 4$ or more.

Levels of *SNB* and *OsIDS1* did not differ between 'Dongjin' and 'Kitaake' rice. The latter is an early-flowering cultivar and carries mutations in *Ghd7* and *PRR37* (Kim et al. [2013]). Therefore, this demonstrated that the AP2 genes are not likely influenced by Ghd7 and PRR37. In addition, AP2 expression was not significantly altered in mutants defective in regulatory genes such as *Ghd7,OsGI, Ehd1, OsMADS50, OsMADS51, OsId1, OsTrx1*, and *OsVIL2*, all of which control flowering time (Additional file 1: Figure S5). However, expression of *SNB* and *OsIDS1* was affected in the *oscol4*and *hd1* mutants (Figure 4A and B; Additional file 1: Figure S5A). This suggested that expression of AP2 genes is positively controlled by *OsCOL4* and *Hd1*.

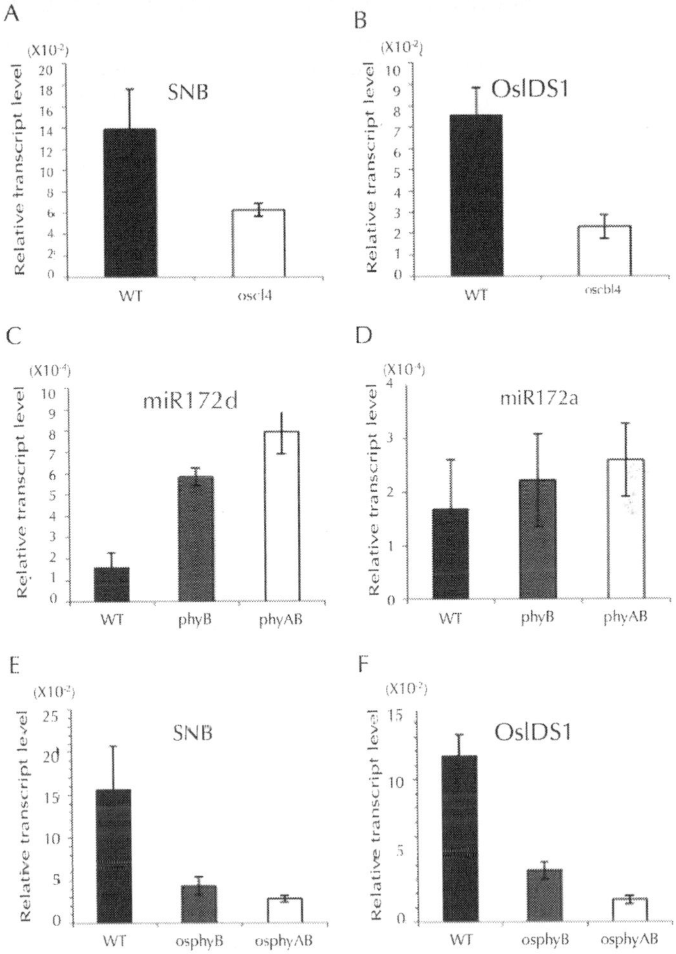

Figure 4: Expression of*miR172*s and AP2 genes in*oscol4*,*osphyB*, and*osphyA osphyB*mutants. (A, B) Transcript levels for *SNB*(A) and*OsIDS1*(B) in WT (black bars) and *oscol4* (open bars) plants under LD.(C, D) Transcript levels for *miR172d*(C) and *miR172a*(D) in WT (black bars), *osphyB* (gray bars), and *osphyA osphyB* (open bars) plants under LD. (E, F) Transcript levels for *SNB*(E) and *OsIDS1*(F) in WT (black bars), *osphyB* (gray bars), and *osphyA osphyB* (open bars) plants under LD. Middle regions of fully emerged uppermost leaves were sampled at 35 DAG. Expression was monitored at ZT 11 h for *miR172*s and at ZT 2 h for *SNB* and *OsIDS1*. Y-axis, relative transcript level compared with *OsUbi1*. Error bars indicate standard deviations;*n* = 4 or more.

miR172d Induces Flowering Time by Suppressing AP2 Genes

The transcripts of *SNB* and *OsIDS1* carry a *miR172* target site (Lee and An [2012]). To examine whether *miR172* controls flowering time, we generated six transgenic rice plants that over-express*miR172d* (Figure 5A). In all plants, transcripts were increased (Figure 5B) and the time to flowering was shortened by 9 to 30 d (Figure 5C). Levels of flowering activators *Ehd1* and *Hd3a* were also substantially higher in the *miR172d* OX lines (Figure 5D and E) while expression of *SNB* and *OsIDS1*was decreased in those lines (Figure 5F and G). These results suggested that *miR172d* induces flowering by reducing AP2 gene expression.

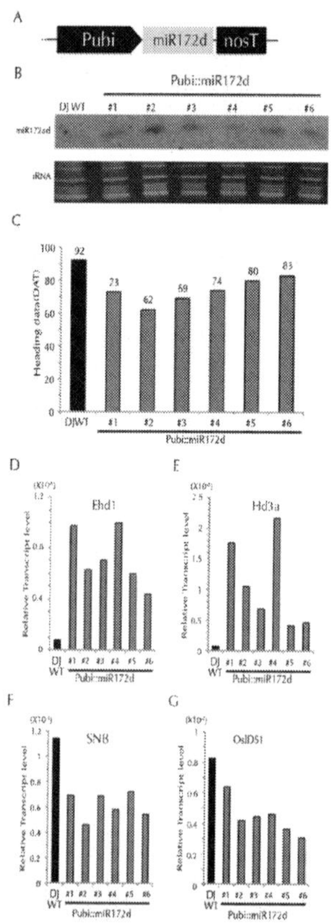

Figure 5: Phenotypes of*miR172d*OX plants. (A) Scheme of *miR172d*OX construct. Spanning region of *miR172d* was sub-cloned into pGA1611 vector between maize *ubiquitin* promoter (*Pubi*) and *nopaline synthase*terminator

(*nos*T). (B) Northern blot analysis of expression levels of mature *miR172* in WT and *miR172d* OX plants (top). Ethidium bromide-stained ribosomal RNA was used as loading control (bottom). (C)Flowering time for 6 *miR172d* OX plants compared with WT when grown in greenhouse. Days-to-heading was scored when first panicle bolted. (D-G)Expression levels of *Ehd1*(D), *Hd3a*(E), *SNB*(F), and *OsIDS1*(G) in WT and *miR172d* OX plants.

To confirm that *miR172* controls flowering time via the AP2 genes, we constructed the *miR172*-resistant form of *SNB* (r*SNB*) by changing the *miR172* target site CTGCAGCATCATCAGGATTCT to CTGCAGCAATGTCCGGATTCT (Figure 6A). Of the six transgenic rice plants carrying this r*SNB*construct, five lines (#1, 3, 4, 5, and 6) expressed the transgene at substantially high levels (Figure 6B). The r*SNB* transcript can be distinguished from the endogenous *SNB* transcript due to the restriction enzyme site *ACC*III that is generated in the r*SNB* OX construct. Here, the RT-PCR products of those r*SNB* OX plants were digested with *ACC*III, supporting our findings that most of the *SNB* transcripts in the OX plants were r*SNB* (Additional file 1: Figure S7). These r*SNB* OX plants eventually died without having flowered, even after several months of growth (Figure 6C). Moreover, transgenic plants over-expressing the *miR172*-resistant form of *OsIDS1* (r*OsIDS1*) did not produce any flowers for more than one year (Additional file 1: Figure S6). These observations further demonstrated that *miR172*s induce flowering time by suppressing AP2 transcripts in rice.

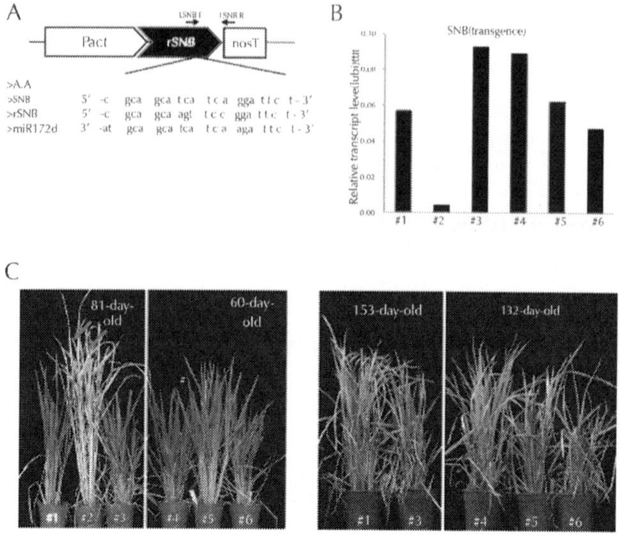

Figure 6: Phenotypes of*miR172*-resistant*SNB*OX plants. (A) Scheme of *miR172*-resistant *SNB* (r*SNB*) construct. Four synonymous mutations (red characters) were introduced into *miR172* target sequences at C-terminus of *SNB* full-length cDNA. Primers t*SNB* F and R were used. Underlined sequence (*tccgga*) indicates restriction enzyme site by *ACC*III. P*act*, *actin* promoter; *nos*T, *nopaline synthase* terminator. (B) Expression level of *SNB* in r*SNB*OX plants. (C) Phenotypes of r*SNB* OX plants grown under SD. Bolted panicles are shown by red arrows.

The Late-Flowering Phenotype of RSNB Ox Plants is rescued by Overexpression of Ehd1

Because *Ehd1* transcripts were reduced in *SNB* OX plants, we postulated that the AP2 gene functions upstream of *Ehd1*. To confirm this hypothesis genetically, we generated transgenic plants over-expressing both r*SNB* and *Ehd1* (Figure 7A). It was previously reported that overexpression of*Ehd1* causes early flowering (Osugi et al. [2011]), and we also observed early-flowering phenotypes of *Ehd1* OX plants (Additional file 1: Figure S8). Thus, if *SNB* functions downstream of *Ehd1*, we would expect that transgenic plants expressing both genes do not flower early. Instead, those plants did flower early and their morphology was similar to plants that over-express only *Ehd1*(Figure 7B and C). This observation supports that *SNB* inhibits flowering time by suppressing *Ehd1*.

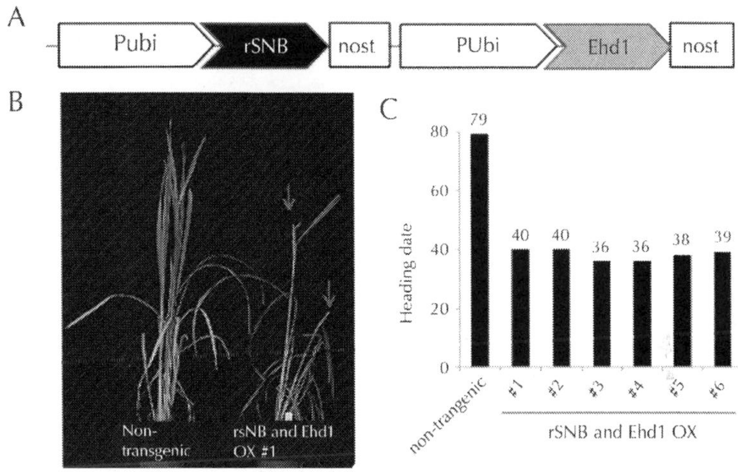

Figure 7: Phenotype of transgenic plant overexpressing both r*SNB*and*Ehd1*. (A) Diagram of *miR172*-resistant *SNB* (r*SNB*) and *Ehd1*double OX construct. p*Ubi*, maize *ubiquitin* promoter; *nos*T, *nos*terminator. (B) Phenotypes of r*SNB* and *Ehd1* OX #1 plants. Photograph was taken at 42 days after transplanting. Red arrows indicate emerged panicles. (C) Heading date of r*SNB* and *Ehd1* OX primary transgenic plants grown in greenhouse.

miR172a and miR172d are Negatively Regulated by Phytochromes in Rice

The level of *miR172* is increased in *phyB* mutants of *Arabidopsis* (Jung et al. [2007]). To study whether *miR172* in rice is also regulated by phytochrome, we measured the levels of pri-*miR172a*and pri-*miR172d* in mutants (Figure 4). This analysis revealed that pri-*miR172a* and pri-*miR172d*transcripts were elevated in the *osphyB* mutant and further increased in the *osphyA osphyB* double mutant (Figure 4C and D). This indicated that *miR172a* and *miR172d* are negatively regulated by phytochrome. Transcript levels of *SNB* and *OsIDS1* were also significantly reduced in the *osphyB*single and the *osphyA osphyB* double mutants (Figure 4E and F). These observations suggested that phytochromes can inhibit flowering time by suppressing *miR172a* and *miR172d*, which interfere with *SNB* and *OsIDS1* expression (Figure 8).

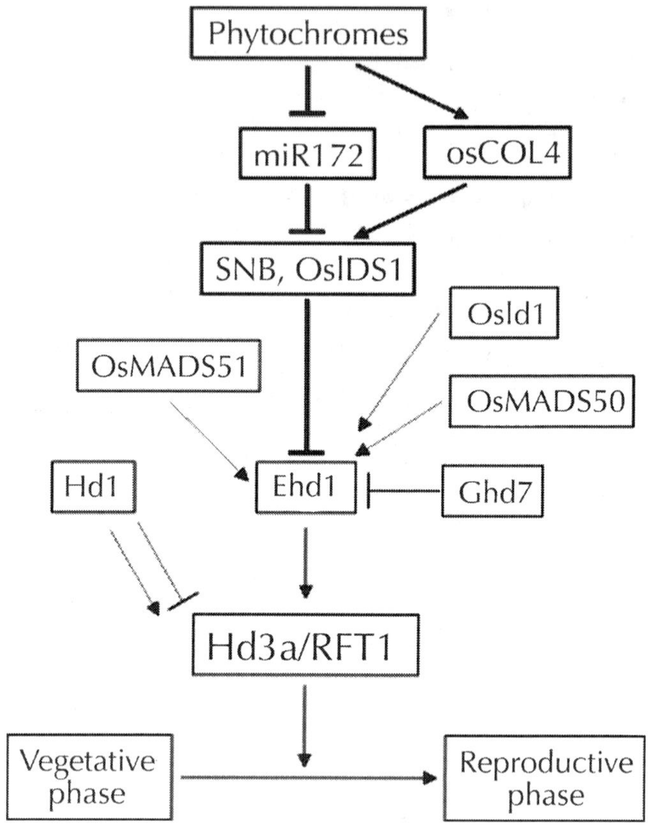

Figure 8: Pathway for control of flowering time via*miR172d*.Phytochromes are negative regulators of *miR172d* that suppress AP2 genes *SNB* and *OsIDS1*, both of which are negative regulators of *Ehd1*.

DISCUSSION

We have demonstrated that *miR172d* induces flowering by suppressing *SNB* and *OsIDS1*. Overexpression of the AP2 genes delays flowering more significantly under SD, the permissive condition. This phenotype is similar to that of *Arabidopsis SMZ*-OX plants (Mathieu et al. [2009]). Overexpression of *SMZ* causes a significant delay in flowers under LD, the permissive condition for that species, but not under SD (Mathieu et al. [2009]). These reports provide evidence that AP2 genes preferentially block the floral transition under inductive conditions.

Although flowering is inhibited under LD, rice plants do flower eventually under normally suppressive conditions, albeit after a certain period of vegetative growth. We showed here that *miR172d* levels were increased in the leaves as a plant aged and that flowering was induced when that gene was over-expressed. This indicated that the microRNA plays a role in the developmental control of flowering time in rice, as also observed from *Arabidopsis* (Jung et al. [2007]).

SNB and *OsIDS1* are flowering repressors that are highly expressed in young leaves. However, their expression gradually declined to minimal levels at 35 DAG when *miR172* transcripts started to increase. Expression of the florigens and *Ehd1* began at 60 DAG while that of the AP2 genes remained low. This indicated that the latter are major suppressors of the downstream floral signals.

Ghd7 also functions to repress *Ehd1*, but its transcription peaks at 2 to 3 weeks before declining to a low level. This occurs much earlier than when the florigen genes are expressed (Matsubara et al.[2011]; Kim et al. [2013]). *OsCOL4* is another regulator that suppresses *Ehd1* and the florigen genes. In particular, *OsCOL4* expression is maintained at a high level in the early vegetative stages but decreases when *Ehd1* expression begins (Lee et al. [2010]). This temporal expression pattern is similar to *SNB* and *OsIDS1* except that *OsCOL4* is reduced later than the AP2 genes. Therefore, these results suggest that the AP2 genes as well as *Ghd7* and *OsCOL4* coordinately suppress flowering time.

Except in *oscol4* and *hd1* plants, *AP2* expression is not significantly altered in most flowering mutants, thereby implying that those genes are controlled by *OsCOL4* and *Hd1*, but not by other flowering time regulators. We have previously reported that OsCOL4 is a constitutive repressor of*Ehd1* (Lee et al. [2010]). Because this is also true of AP2 genes, it is likely that OsCOL4 suppresses*Ehd1* by inducing AP2 expression. Hd1 also suppresses flowering time via repressing *Hd3a* and*RFT1* under LD conditions (Yano et al. [2000]; Hayama et al. [2003]). These indicate that the AP2 genes and *OsCOL4* co-operatively suppress flowering time. We observed that *miR172d* expression is negatively affected by phytochrome activity. Considering that AP2 genes are controlled by miR172, we might conclude that phytochromes support vegetative growth by maintaining AP2 expression. Both *Ghd7* and *OsCOL4* are positively modulated by *OsPhyB*(Lee et al. [2010]; Osugi et al. [2011]). Therefore, it is apparent that phytochromes influence these flowering regulators collectively.

In plants, the *miR172/AP2* module is inversely correlated with the *miR156*/SPL module (Aukerman and Sakai [2003]; [2005]; Wu and Poethig [2006]; Chuck et al. [2007]; Poethig [2009]). In rice,*miR156* genes are predominantly expressed in the young shoots, etiolated shoots, and seedling leaves (Xie et al. [2006]) and its target *OsSPL* genes are expressed high in young panicles (Xie et al.[2006]). Overexpression of *miR156* OX causes various phenotypes such as increase of tiller numbers, late flowering, dwarfism, and decrease of spikelet numbers (Xie et al. [2006]). In maize, overexpression of the maize *miR156* prevents flowering and increases starch content (Chuck et al.[2007]; [2011]). This late-flowering phenotype is similar to that of *SNB* OX or *OsIDS1* OX plants. Overexpression of the maize *miR156* also causes increase of biomass and tiller number in maize and also in *Arabidopsis*, *Brachypodium* and switchgrass (Chuck et al. [2011]). These phenotypes are similar to those of *OsmiR156* OX (Xie et al. [2006]; Chuck et al. [2011]).

CONCLUSIONS

We demonstrated that over-expressions of *OsIDS1* and *SNB* inhibited flowering time in rice by suppressing expression of *Ehd1*. In addition, expressions of the AP2 genes were repressed by*miR172* and the later was increased in the *osphyB* and *osphyA osphyB* mutants. Based on these findings, we concluded that *miR172* induced flowering by suppressing the AP2 genes and the microRNA gene was inhibited by phytochromes. The facultative LD-flowering phenotype of rice can be explained in part by the miR172-AP2 pathway.

METHODS

Plant Material and Growth Conditions

We have previously developed T-DNA-tagging lines in *Oryza sativa* japonica cv. Dongjin (Jeon et al.[2000]). The flanking sequences were determined via inverse PCR (An et al. [2003]; Ryu et al.[2004]). T-DNA insertional mutants *osphyA-2, osphyB-2, osmads50-1, osmads51-1, hd1-1,oscol4-2, osvil2-1,* and *ostrx1-1* were reported earlier (Lee et al. [2004]; Jeong et al. [2007]; Kim et al. [2007]; Ryu et al. [2009]; Lee et al. [2010]; Yang et al. [2013]; Choi et al. [2014]). RNAi-suppressed plants of *Ehd1* RNAi and *OsId1* RNAi have been described previously (Kim et al. [2007]; Park et al. [2008]). A near isogenic line (NIL) that carries the *ghd7* allele was presented by Kim et al. ([2013]). Seeds were germinated on a half-strength Murashige and Skoog medium containing 3% sucrose. They were incubated for one week at 28°C under constitutive light as previously reported (Yi and An [2013]). Seedlings were transplanted into soil and cultured in growth chambers under either SD (12 h light at 28°C, humidity 70%; 12 h dark at 25°C, humidity 50%) or LD (14.5 h light at 28°C, humidity 70%; 9.5 h dark at 25°C, humidity 50%) as previously reported (Yi and An[2013]).

Construction of *Osphya Osphyb* Double Mutants

The *osphyA osphyB* double mutants were obtained by crossing *osphyA-2* and *osphyB-2* single mutants. This *osphyA-2* mutant is a null allele generated by a T-DNA insertion in the fourth exon of*OsphyA*. The *osphyB-2* mutant is also a null allele produced by inserting T-DNA in the third intron of*OsphyB* (Jeong et al. [2007]). Afterward, F2 segregants were genotyped by PCR (35 cycles of 95°C for 15 s, 55°C for 30 s, and 72°C for 60 s), using a combination of gene-specific primers and the T-DNA primer 5′-TTGGGGTTTCTACAGGACGTAAC-3′. Those gene-specific primers were 5′-CAGGGAAAAGGGATTAGAGT-3′ and5′-AGTGGACTCGGGTTAACTTT-3′for*osphyA-2*, and 5′-AGTTAAGACGGAGCGACATA-3′ and 5′-GTAAGCGATCAGTTTGTGGT-3′ for *osphyB-2*. For further analyses, we selected F2 progeny that had homozygous mutant alleles for both genes.

ABSTRACT

Background

Plants have evolved a sensitive defense response system that detects and recognizes various pathogen-associated molecular patterns (PAMPs) (e.g. flagellin) and induces immune responses to protect against invasion. Transcriptional responses in rice to PAMPs produced by *Xanthomonas oryzae* pv. *oryzae* (Xoo), the bacterial blight pathogen, have not yet been defined.

Results

We characterized transcriptomic responses in rice inoculated with the wildtype (WT) Xoo and flagellin-deficient mutant Δ*fliC* through RNA-seq analysis. Digital gene expression (DGE) analysis based on Solexa/Illumina sequencing was used to investigate transcriptomic responses in 30 day-old seedlings of rice (*Oryza sativa* L. cv. Nipponbare). 1,680 genes were differentially-expressed (DEGs) in rice inoculated with WT relative to Δ*fliC*; among which 1,159 genes were up-regulated and 521 were down-regulated. Expression patterns of 12 randomly-selected DEGs assayed by quantitative real time PCR (qRT-PCR) were similar to those detected by DGE analyses, confirming reliability of the DGE data. Functional annotations revealed the up-regulated DEGs are involved in the cell wall, lipid and secondary metabolism, defense response and hormone signaling, whereas the down-regulated ones are associated with photosynthesis. Moreover, 57 and 21 specifically expressed genes were found after WT and Δ*fliC* treatments, respectively.

Conclusions

DEGs were identified in rice inoculated with WT Xoo relative to Δ*fliC*. These genes were predicted to function in multiple biological processes, including the defense response and photosynthesis in rice. This study provided additional insights into molecular basis of rice response to bacterial infection and revealed potential functions of bacterial flagellin in the rice-Xoo interactions.

BACKGROUND

Plants, unlike mammals, do not have an adaptive immunity mechanism, and largely rely on the innate immune system to resist pathogen infection (Jones and Dangl [2006]). Pattern-triggered immunity (PTI) and effector-triggered immunity (ETI) are two layers of innate immune system in plant (Chisholm et al. [2006]; Tsuda and Katagiri [2010]). Pathogen-associated molecular patterns (PAMPs) are conserved and indispensable molecules, which include bacterial flagellin, elongation factor-Tu (EF-Tu), fungal glucans, chitin, etc. (Zhang and Zhou [2010]). PTI is activated through recognition of PAMPs by pattern recognition receptors (PRRs) located on the plant cell surface; the resulting basal defense response prevents potential pathogens from colonizing on plant tissues (Chakravarthy et al. [2009]; Chujo et al. [2013]). PTI is characterized by induction of reactive oxygen species (ROS), ion fluxes, callose deposition in the cell wall and other innate immune responses (Tsuda et al. [2008]; Boller and Felix [2009]). During the evolution of plant-pathogen interactions, some pathogens acquired the ability to suppress PTI by delivering effectors to plant cells, thus causing disease (Li et al. [2005]; Chisholm et al. [2006]). Then plants developed resistance (R) proteins to recognize effectors, leading to activation of effector-triggered immunity (ETI) (Jones and Dangl [2006]). Pathogenesis related proteins (PRs) and WRKY transcription factors (WRKY TFs) are two kinds of important proteins involved in defense to pathogens in both PTI and ETI (Chujo et al. [2013]).

Flagellin, the main component of a bacterial flagellum, was first identified as an elicitor that induces defense responses in tomato and *Arabidopsis thaliana* (Felix et al. [1999]). Flg22, a 22 amino acid, highly conserved domain within the N terminus of flagellin, caused these defense responses and a strong inhibition of growth in *A. thaliana* seedlings (Gomez-Gomez et al. [1999]). FLAGELLIN SENSITIVE2 (FLS2), the receptor that recognizes flagellin in *A. thaliana*, shares structural and functional homologies with known plant resistance proteins (Gomez-Gomez and Boller [2000]). Flagellin and flg22 are recognized and bound to the extracelluar domain of FLS2. This induces FLS2 heteromerization with BRASSINOSTEROID INSENSITIVE 1-associate kinase 1 (BAK1) and activates signaling cascades that finally culminate in the induction of defense responses

(Heese et al.[2007]; Ali and Reddy [2008]; Sun et al. [2013]). Therefore, FLS2 determined the specificity of flagellin perception in *A. thaliana* (Chinchilla et al. [2006]). OsFLS2, the ortholog of FLS2 in rice, also mediates immune responses induced by flagellin (Takai et al. [2008]).

Different responses of rice cells to flagellins from compatible or incompatible bacterial strains have been observed. For example, flagellin purified from an incompatible strain of *Acidovorax avenae*(N1141), induced the rapid generation of H_2O_2 accompanying hypersensitive cell death and the expression of *PAL, Cht-1*,and *PBZ1* in cultured rice cells, whereas the flagellin from the compatible strain K1 did not (Che et al. [2000]; Tanaka et al. [2003]). Introduction of N1141 flagellin gene into rice also trigged immune responses (Takakura et al. [2008]). However, flg22 did not induce defense responses in rice, suggesting that the recognition mechanism for flagellin might be different between rice and dicotyledonous plants, such as Arabidopsis and tomato (Felix et al.[1999]; Takai et al. [2007]). Interestingly, when expressed and purified from *Escherichia coli,* both types of flagellins induced H_2O_2 generation in rice. In addition, a deglycosylated flagellin from the compatible strain induced the same immune responses as the flagellin of incompatible strain did (Hirai et al. [2011]). These studies suggested that post-translational modifications of flagellins might be associated with the specific induction of immune responses in rice. However, the molecular mechanism of different immune responses of induced by different flagellins remains largely unknown.

Xanthomonas oryzae pv. *oryzae* (Xoo), causes bacterial blight disease and can result in serious yield loss in rice production. The Xoo-rice interaction has been studied as a model system to understand the molecular mechanisms of disease resistance responses in monocotyledonous plants (Song et al. [1995]; Ronald [1997]; Martin et al. [2003]). Microarray studies revealed that several signaling components, membrane bound receptor kinases and disease-resistant proteins were significantly induced after Xoo inoculation (Narsai et al. [2013]). However, the function of flagellin in Xoo-rice interaction has never been studied. In a previous study, we generated a mutant of Xoo that was deleted in the flagellin gene *fliC* (Δ*fliC*); the Δ*fliC* mutant was not motile, and caused more disease (increased lesion lengths) in rice leaves relative to the wildtype (WT) strain (Tian et al. [2014]), suggesting that differential rice responses to the WT and Δ*fliC*inoculations exist.

In this study, digital gene expression (DGE) based on Solexa/Illumina sequencing, was applied to identify differentially-expressed genes (DEGs) in rice inoculated with WT Xoo relative to the Δ*fliC*mutant. We identified 1,680 DEGs involved in cell wall and lipid synthesis, secondary metabolism, photosynthesis, defense response, and hormone signaling pathways.

RESULTS

DGE sequencing in rice leaves inoculated with WT Xoo or Δ*fliC*

For DGE sequencing, RNAs from rice leaves inoculated with the WT or Δ*fliC* mutant were extracted to prepare two cDNA libraries for RNA-seq analysis. After filtering to remove reads containing the adapter and poly-N or low quality reads, 36,187,662 and 38,239,937 clean reads remained in the WT and Δ*fliC* libraries, respectively. The sequencing depth was sufficient for the transcriptome coverage in rice (Table 1).

Table: Summary of sequencing data

Sample	Raw tag	Clean tag		GC (%)	Tags mapped to gene	
		TN	TP (%)		TN	TP (%)
WT	40,067,282	38,239,937	95.44	54.93	32,279,312	84.41
fliC	37,643,149	36,187,662	96.13	52.53	30,423,983	84.07

TN: total number; TP: total percentage.

Yu *et al.*

Yu *et al.* *Rice* **2014 7**:20 doi:10.1186/s12284-014-0020-7

A total of 30,423,983 (84.07%) and 32,279,312 (84.41%) clean reads in the two libraries were mapped to the reference genome of rice using bowtie software and allowing a 2-bp mismatch. The GC contents were 52.53% and 54.93%, respectively (Table 1). Over 80% clean tags per library were mapped to the reference database, showing the DGE data was reliable and sufficient for subsequent bioinformatics analysis of gene expression.

The raw sequencing data obtained in this work have been deposited in NCBI's Sequence Read Archive (SRA) and are accessible through SRA Series accession number PRJNA238154, and the accession numbers of WT and Δ*fliC* libraries are SRR1168425 and SRR1168426, respectively

Identification of DEGs in Rice Inoculated with Wt Xoo Relative to Δ*Flic*

To compare differential gene transcription in WT- and Δ*fliC*-treated rice, two transcriptome profiles were analyzed. As described in materials and methods, the number of clean reads in each library was normalized to the number of transcripts per kilobase of exon model per million mapped reads (FPKM) to obtain the normalized gene expression level, then the relative expression levels were calculated by taking Δ*fliC* treatment as the control. Among the 1,680 DEGs identified, 1,159 were up-regulated and 521 were down-regulated in the WT relative to Δ*fliC* treatments (Figure 1). In addition, 57 and 21 DEGs were specifically expressed in WT and Δ*fliC* libraries, respectively (Figure 1B).

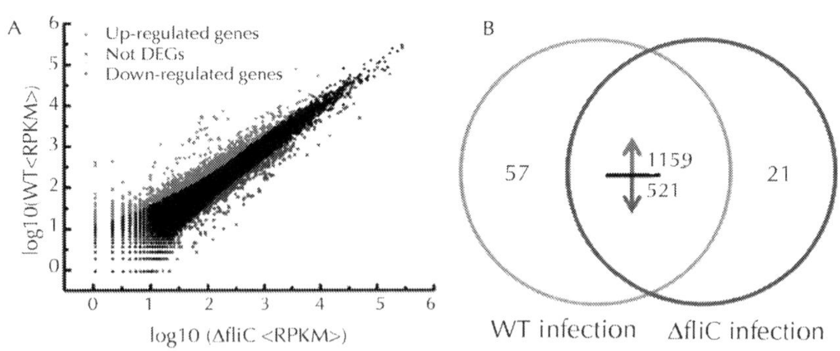

Figure 1: Differentially-expressed genes (DGEs) between the WT and **Δ***fliC*treatments in rice. A, Comparison of gene expression levels between

WT and Δ*fliC* treatments. Red and blue dots represent transcripts more prevalent in WT and Δ*fliC* libraries, respectively. Black dots indicate transcripts that did not change significantly. The parameters "P ≤ 0.05" and "|$\log_2$ Ratio| ≥ 1" were used as the threshold to judge the significance of gene expression difference. B, Changes in gene expression between the WT and Δ*fliC* libraries. Using the Δ*fliC* library as control, red and blue arrows represent genes were up-regulated and down-regulated in WT library, respectively. Data represent the number of DEGs.

To validate the reliability of DGE sequencing, the expression of 12 randomly-selected DEGs were revealed by quantitative real time PCR (qRT-PCR) assays in another independent inoculation experiment. The results showed all of gene expression patterns were consistent to those from the DGE analysis, indicating the DGE data was reliable (Figure 2).

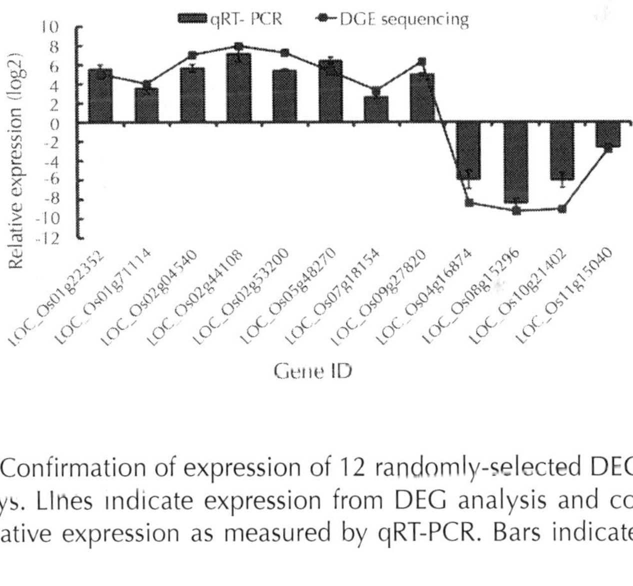

Figure 2: Confirmation of expression of 12 randomly-selected DEGs by qRT-PCR assays. Lines indicate expression from DEG analysis and columns indicate relative expression as measured by qRT-PCR. Bars indicate standard error.

Functional Classifications of DEGs

A total of 1,758 DEGs between WT and Δ*fliC* treatments were categorized into 49 functional groups using GO classifications (Figure 3). In the biological process category, metabolic process was the most dominant group, followed by cellular process, response to stimulus and biosynthetic process. Additionally, there were many DEGs involved in carbohydrate metabolism, photosynthesis and secondary metabolism. Among the cellular components, cell was the largest group, followed by cell part and cytoplasm. Catalytic activity was the largest group in molecular function, followed by hydrolase activity, and transporter activity. These results indicate that the bacterial flagellin influences expression of genes involved in multiple physiological and biochemical activities in rice.

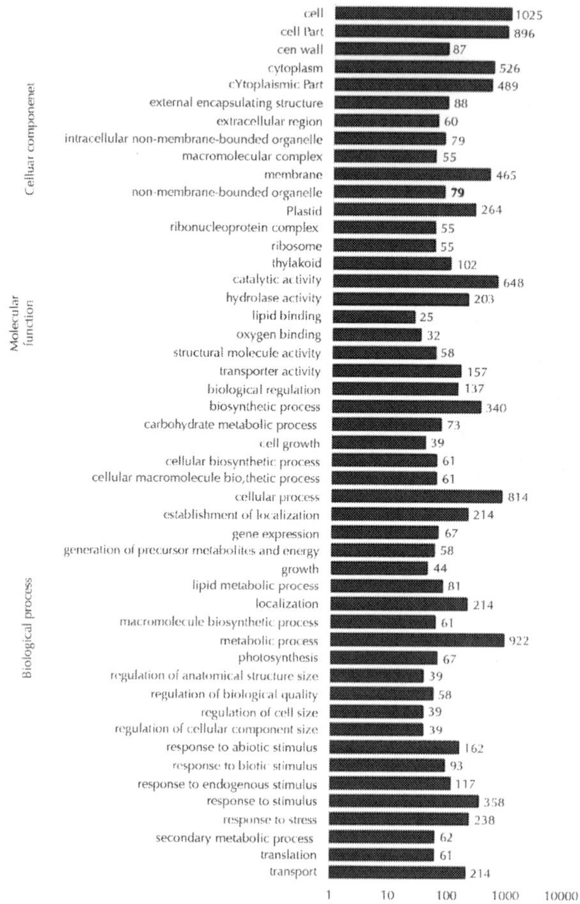

Figure 3: Histogram showing gene ontology functional analysis of DEGs. The y-axis and X-axis indicated the names of clusters and the number of DEGs in clusters, respectively.

Metabolism Pathway Analysis of DEGs

To further reveal the potential functions of DEGs, an overview of the metabolic processes in rice transcriptionally affected by WT and Δ*fliC* infection was obtained by using MapMan (Thimm et al. [2004]). Cell wall, lipids metabolism, secondary metabolism, amino acid synthesis and degradation, photosynthesis, and other metabolic processes were detected (Figure 4). The DEGs associated with photosynthesis showed significant expression differences between WT- and Δ*fliC*-treated rice. Among them, genes involved in light reactions, Calvin cycle and ATP synthesis were down-regulated, while those involved in starch and sucrose degradation were up-regulated (Additional file 1: Table S5). These results suggest that bacterial flagellin suppresses photosynthetic assimilation in rice.

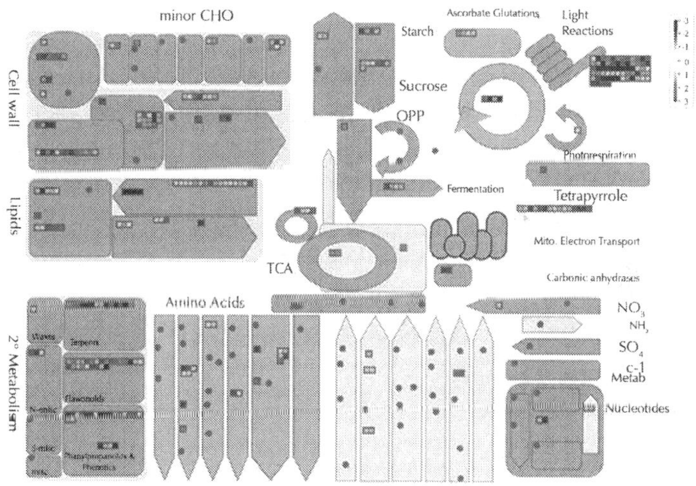

Figure.4: Metabolism overview between WT and **Δ***fliC*treatments by MapMan. 1,680 DGEs (fold changes ≥ 2, p-value ≤ 0.05) were imported into MapMan. DGEs up- and down-regulated are represented with red and blue squares, respectively. The grey circles means no differentially-expressed genes involved in this process. Values are log_2-transformed fold changes.

Lipid metabolism has an important role in defense response to pathogen infection by providing the energy. Five phospholipid

modification, RNA transcription, signaling, development, transport, etc. (Figure 6and Additional file 5: Table S6). In addition, 21 SEGs in the Δ*fliC*-treated library were identified to be involved in secondary metabolism, RNA transcription, protein degradation and modification, signaling and transport (Figure 7 and Additional file 6: Table S7).

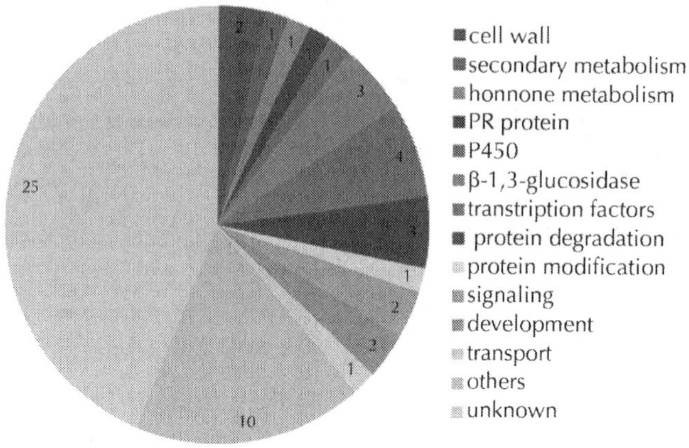

Figure 6: Genes specifically expressed in the WT library. Data represent the number of DEGs.

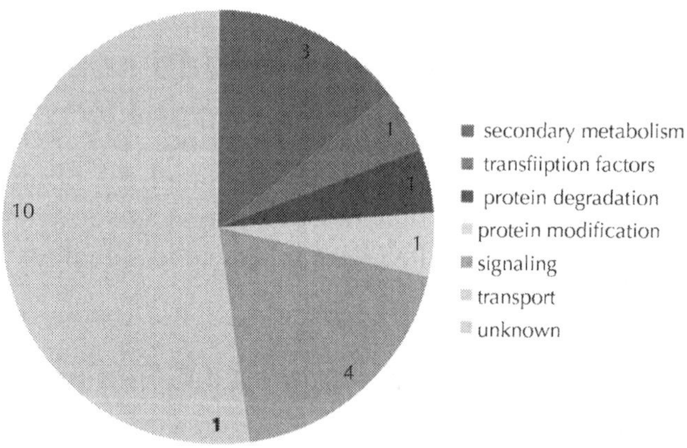

Figure 7: Genes specifically expressed in the **Δ***fliC*library. Data represent the number of DEGs.

DISCUSSION

Flagellin, the main component of flagella, not only plays an important role in bacterial motility, but it is also important in plant PTI (Lee et al. [2003]). Flagellin from avirulent but not virulent strains of the rice pathogen *Acidovorax avenae* induced hypersensitive cell death in cultured rice cells (Che et al. [2000]); the specificity of this induction of immune responses was attributed to flagellin glycosylation (Taguchi et al. [2010]; Hirai et al. [2011]), and suggested that the function of bacterial flagellin in PTI in rice relied on its origin and post-translational modification. We have shown that a *fliC* deletion mutant of Xoo causes increased disease (longer lesion lengths) in rice leaves relative to WT Xoo (Tian et al. [2014]). In this study, we identified DEGs by comparing transcriptomic responses in rice inoculated with the WT Xoo and Δ*fliC*. The induction of *fliC* gene expression of Xoo during bacterial growth in rice indicated the expression levels of these DEGs were changed with flagellated Xoo invasion (Additional file 7: Figure S1). The putative functions of these DEGs revealed multiple biological processes, including cell wall synthesis, photosynthesis, defense response and hormone signaling, etc. are involved in rice responses to Xoo flagellin, and provide additional insights into the roles of bacterial flagellin played in the Xoo-rice interactions.

Expression of genes involved in cell wall formation and degradation were influenced by bacterial infection. Microbial pathogens enhance infection by degrading plant cell walls using cell-wall-degrading enzymes. On the other hand, cell wall degradation products produced by microbial activity can induce innate immune responses in plants, including callose deposition and programmed cell death (Jha et al. [2007]; Aparna et al. [2009]). In addition, some genes involved in cell wall modification play an important role in basal disease resistance (An et al. [2008]). Our current finding that up-regulation of DEGs associated with cell wall precursor synthesis, cell wall degradation and modification (Figure 4 and Additional file 2: Table S2) is consistent with other reports (Narsai et al. [2013]; Socquet-Juglard et al. [2013]).

PRs are synthesized during basal defense responses and enhanced disease resistance to pathogens in plants. Flagellin induces the expression of PR genes in *Arabidopsis* (Gomez-Gomez et al. [1999]). WRKY TFs form a large family of plant-specific transcription factors and can be positively or negatively regulated during defense responses (Chujo et al. [2013]). Constitutive expression of WRKY18 enhanced resistance to *Pseudomonas syringae* in *Arabidopsis*, while its co-expression with WRKY40 made plants more susceptible to *P. syringae* (Xu et al. [2006]; Schon et al. [2013]). Our results revealed seven up-regulated PRs involved in basal disease resistance, and down-regulated genes for WRKY18 and WRKY40 (Figure 5 and Additional file 3: Table S3), demonstrating these particular genes might play important roles in rice-Xoo interactions.

Pathogen infection usually results in the significant decreases of photosynthesis, and in turn, limits the availability of nutrient sources for pathogen (Berger et al. [2007]). For example, the photosynthesis was decreased in *Arabidopsis* leaves after challenged with *P. syringae* (Bonfig et al.[2006]). Genes encoding photosynthesis components were repressed by Xoo infection in rice (Narsai et al. [2013]). Activation of defense by PAMPs leads to rapid decrease in nonphotochemical quenching (Gohre et al. [2012]). Our data showed genes encoding proteins involved in light reactions, Calvin cycle and ATP synthesis were repressed in rice after treatment with WT Xoo relative to the mutant (Figure 4 and Additional file 4: Table S4), implying the bacterial flagellin might function in suppression of photosynthesis of rice.

Roles of salicylic acid (SA) and ethylene (ET)/jasmonic acid (JA)-mediated signaling pathways in resistance to pathogens have been well studied (Robert-Seilaniantz et al. [2011]). SA synthesis was accelerated by pathogen infection, and enhanced the induction of defense gene transcripts, H_2O_2 accumulation and hypersensitive cell death in plants (Gaffney et al. [1993]; Shirasu et al.[1997]), and ET/JA signaling pathway was required for resistance to necrotrophic pathogens(Glazebrook [2005]; Denance et al. [2013]). In addition, ET synthesis was induced by the PAMP flg22 (Felix et al. [1999]). In our study, several genes encoding 1-aminocyclopropane-1-carboxylate oxidase (ACO) and 12-oxophytodienoate reductase (OPR) were up-regulated (Figure 5and Additional file 1: Table S5); these are key enzymes involved in ET and JA synthesis pathway, respectively (Vriezen et al. [1999]; Schaller et al. [2000]). These results indicated ET/JA synthesis were induced by Xoo flagellin in rice. Interestingly, seven genes encoding salicylic acid carboxyl methyltransferase (SAMT), catalyzing SA to methyl SA or dimethyl SA (Fukami et al. [2002]), were down-regulated. We suggest that SA accumulation in rice is induced by Xoo flagellin. Genes involved in auxin and brassinosteroid signaling pathways were also identified, but the relationship of these hormones and plant defense response is largely unknown.

CONCLUSION

A number of rice genes differentially expressed after inoculation with WT Xoo relative to Δ*fliC* was identified by DGE analysis, some of which were validated by qRT-PCR assays for their expression patterns. These DEGs were involved in the multiple different biological functions, including induction of cell wall and lipid synthesis, secondary metabolism, defense responses, hormone signaling and suppression of photosynthesis. This study provided additional insights into molecular basis of rice response to bacterial infection and revealed potential functions of bacterial flagellin in the rice-Xoo interactions. Functional characterization is further required for these candidate genes of rice via the transgenic approaches including the gene over-expression and/or silencing analysis.

METHODS

Plant Materials and Bacterial Inoculation Treatment

Rice seeds (*Oryza sativa* L. cv. Nipponbare) were germinated for 2 days in water at 37°C. Individual germinated seeds were planted in a small pots containing vermiculite, the pots were placed in a tray that contained 1 L Hogland's solution, which was a mixture of 10 mM KH_2PO_4, 2 mM $MgSO_4$, 1 mM $CaCl_2$, 0.1 mM Fe-EDTA, 50 μM H_3BO_4, 12 μM $MnSO_4$, 1 μM $ZnCl_2$, 1 μM $CuSO_4$, 0.2 μM Na_2MoO_4 and 3 mM KNO_3. The pH of the nutrient solution was adjusted to 5.5-6 using 50% phosphoric acid, and the solution was replaced every two days. Growth conditions were 16 h light (29°C) and 8 h of dark (23°C). Xoo strain $PXO99^A$ (Hopkins et al. [1992]) is a virulent strain to Nipponbare. Δ*fliC*, which is a *fliC* deletion mutant derived from $PXO99^A$, was generated by marker exchange strategy as previous described (Tian et al. [2014]). PXO99^A and Δ*fliC* were grown for 72 h at 28°C in M210 media (Yang et al. [2012]), then the cells were collected by centrifugation and resuspended in d_2H_2O at an OD_{600} of 0.8 (approximately 1.0×10^8 CFU/mL). Thirty day-old seedlings of rice were inoculated with PXO99^A and Δ*fliC* by the leaf-clipping method (Chen et al.[2002]). Five centimeters tip of inoculated rice leaves were collected at 12 h after inoculation, frozen in liquid nitrogen, and stored at −80°C.

In order to detect the *fliC* expression of during Xoo growth in rice, we collected the rice leaves, which were inoculated with PXO99^A at 7 d after inoculation. Then total RNA was extracted using Trizol reagent (Invitrogen, USA), and treated with DNase. The first strand cDNA fragment of *fliC*was synthesized using Superscript III reverse transcriptase (Invitrogen, USA). Specific primer sets, fliC-F (5′ –CCGAGCGTTTCACTACCCA– 3′) and fliC-R (5′ –ATCCTTGAACGACAGGCTGAT– 3′) were designed based on the sequence of *fliC*. PCR was performed with one denaturation cycle of 5 min at 98°C and 32 cycles of 30 s at 95°C, 30 s at 55°C, and 30 s at 72°C, then 10 min at 72°C. Ten microlitres of the PCR product was loaded onto 2% agarose gel. The PCR was performed without Superscript III reverse transcriptase as a negative control.

Solexa/Illumina Sequencing

Total RNA was extracted from rice leaves in WT treatment and Δ*fliC* treatment using Trizol reagent (Invitrogen). According to the manufacturer's instruction of Solexa/Illumina sequencing, mRNA were isolated and enriched from 6 μg total RNA by using the oligo(dT) magnetic beads, then the first- and second-strand cDNA were synthesized by oligo(dT) primer. cDNA were digested with*Nla*III, and the cDNA fragments were ligated with adapter 1, digested with *Mme*I to produce tags with adaptor 1. After removing the magnetic beads, Illumina adaptor 2 was ligated to the 3′ ends of cDNA fragments, and a tag library was formed by different adaptors at both ends of fragments. Lastly, the library fragments were amplified by PCR and purified, and then sequenced via Illumina HiSeq™ 2000.

Identification of Differentially-Expressed Genes

Raw data from Solexa/Illumina sequencing were cleaned by removing reads containing the adapter or poly-N, orlow quality reads. At the same time, the Q20 content and GC content and sequence duplication level of the clean data were calculated. All downstream analyses were based on the clean data with high quality. All clean tags were mapped to reference sequences using bowtie software and allowing 2-bp mismatch. The number of clean tags for each gene was calculated and then normalized to reads per kilobase of exon model per million mapped reads (FPKM). Differentially expressed genes between two samples were identified by tophat and cufflinks softwares (Trapnell et al. [2009]; Roberts et al. [2011]), The $P \leq 0.05$ and the absolute value of $\log_2$ Ratio ≥ 1 were chosen as the threshold to judge the significance of gene expression difference.

Functional Annotation of DEGs

Gene Ontology (GO) (Ashburner et al. [2000]) and MapMan (Thimm et al. [2004]) were used to analysis biological function of DEGs. The GO enrichment analysis was applied to describe product characteristics and reaction network of DEGs. All DEGs were mapped to GO terms in the database (http://bioinfo.cau.edu.cn/agriGO/ *webcite*). GO terms with corrected P value less than 0.05 were considered significantly enriched by differential expressed genes. MapMan software was used to visualize the expression change levels of individual genes in diagrams of metabolic pathways. As previously described (Thimm et al. [2004]), an *Oryza sativa* map was developed and uploaded to MapMan, and the change in expression ratio of each gene was calculated as the $\log_2$-fold change to generate the MapMan experimental file.

Quantitative Real-Time PCR (qRT-PCR) Analysis

Analysis of qRT-PCR were carried out using SYBR Green detection reagents (Quanta Biosciences, USA) in Applied Biosystem's 7500 Sequence Detection System (Applied Biosystems, USA), and the 20 μl PCR reaction contained about 100 ng of cDNA. The reaction mixture was incubated at 95°C for 3 min, and then 40 cycles of 95°C for 10 s, 60°C for 30s, followed by a disassociation stage. Twelve DEGs were randomly selected, and specific primers of these genes were designed with the software Primer Premier 5.0 (PREMIER Biosoft Int., USA) (Additional file 8: Table S1). The rice actin gene was selected as a reference gene in qRT-PCR. The relative expression ratio was calculated using $2^{-\Delta\Delta Ct}$ method (Schmittgen and Livak [2008]). All samples were performed in three biological replicates and triplicate PCR, and the error bars show standard error.

Competing Interests

The authors declare that they have no competing interests.

Author Contributions

Conceived and designed the experiments: CY HMC CYH. Performed the experiments: CY HMC. Analyzed the data: CY HMC FT JEL CYH. Contributed reagents/materials/analysis tools: HMC FT. Wrote the paper: CY HMC FT JEL CYH. All authors read and approved the final manuscript.

ACKNOWLEDGEMENTS

This work was supported by the grants from the National Basic Research Program (973) (Grant#2011CB100700) and National High-Technology Research Program (863) (Grant# 2012AA101504) to C. He.

REFERENCES

1. Ali GS, Reddy A (2008) PAMP-triggered immunity: Early events in the activation of FLAGELLIN SENSITIVE2. Plant Signal Behav 3:423-426
2. An SH, Sohn KH, Choi HW, Hwang IS, Lee SC, Hwang BK (2008) Pepper pectin methylesterase inhibitor protein CaPMEI1 is required for antifungal activity, basal disease resistance and abiotic stress tolerance. Planta 228:61-78

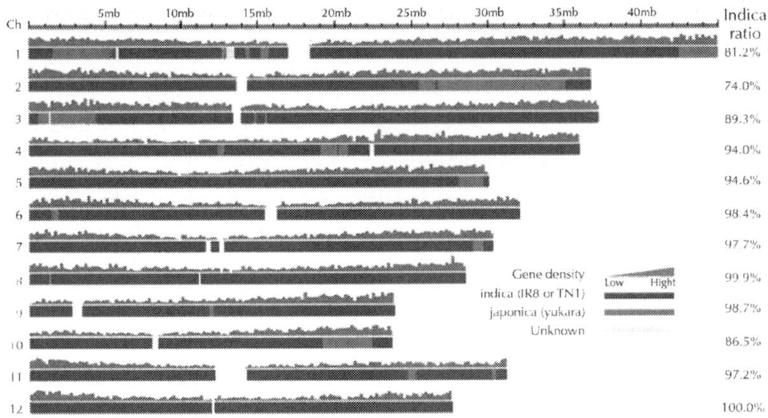

Figure 2: *Indica/japonica*genome organization on the 12 chromosomes of Tongil. Blue indicates the *indica* genome (TN1 and IR8); red indicates the *japonica* genome (Yukara); and yellow indicates a region from an unknown genome. The percentages describe the proportion of*indica* contribution on each chromosome.

Table 2: Determination of theindica/japonicagenome origin of Tongil, based on a window size of 9

Chrom-osome	Pseudom-olecule	indicaregion (bp)	Ratio (%)	***Japonicar-egion* (bp)**	Ratio (%)	Unknown region (bp)	Ratio (%)
1	45,038,604	36,563,905	81.2	7,596,808	16.9	877,891	2.0
2	36,792,247	27,235,850	74.0	9,544,379	25.9	12,018	0.0
3	37,312,367	33,336,733	89.3	3,748,667	10.1	226,967	0.6
4	36,060,865	33,898,364	94.0	2,150,911	6.0	11,590	0.0
5	30,073,438	28,436,341	94.6	1,637,097	5.4		
6	32,124,789	31,619,689	98.4	499,676	1.6	5,424	0.0
7	30,357,780	29,667,148	97.7	690,632	2.3		
8	28,530,027	28,487,631	99.9	333		42,063	0.2
9	23,895,721	23,592,877	98.7	302,844	1.3		
10	23,703,430	20,504,662	86.5	3,198,768	13.5		
11	31,219,694	30,345,040	97.2	846,802	2.7	27,852	0.1
12	27,679,166	27,679,166	100.0				
Tntal	387.788.178	3S1.3R7.4RR	91.R	311.216.917	7.9	1.703.805	0.3

Gene Distribution and Gene Ontology Analysis of Tongil

We analyzed the gene content of Tongil to understand the relationship between the composition of the genome and genes (open reading frames: ORFs), and also to elucidate the distribution of *indica*- and *japonica*-originated genes (alleles) within the Tongil genome. The gene distribution ratio according to *indica* or *japonica* genome composition was similar to the genome distribution ratio (Table 2 and Additional file 5: Table S4). The origins of genes from the *indica* and *japonica* parents were 88.3% and 11.4%, respectively, suggesting that the average gene composition was similar to the genome composition ratio of Tongil, although the distribution of parental origin varied across chromosomes. We performed gene ontology (GO) analysis of the Tongil genome according to three categories to identify biological patterns using a list of genes derived from *indica, japonica,* and unknown genomes: cellular components, molecular functions, and biological processes (Additional file 6: Figure S2; Additional file 7: Figure S3; Additional file 8: Figure S4). The results of GO analysis revealed that the average contribution of the *indica* or *japonica* genome to each GO category was almost identical to the gene and genome distribution ratios. *O. s. indica* and *O. s. japonica* contributed 86.8% and 12.7% of the cellular components, 87.4% and 12.2% of the molecular functions, and 87.3% and 12.2% of the biological processes, respectively, to the Tongil genome. However, in the 'molecular functions' category, all 17 genes related to channel regulator activity were derived from *indica* regions, whereas all adhesion-related genes in the biological processes category were derived solely from *japonica* regions.

Simple Sequence Repeats (SSRs) in the Tongil Genome

A total of 177 distinctive motif families were annotated on the Tongil genome (Additional file 9: Figure S5; Additional file 10: Figure S6). Di-nucleotide repeats were predominant among the classified repeats, and AT/TA repeats were the most abundant motifs in both *indica*- (29.09%) and *japonica*-derived (21.8%) regions within the Tongil genome. The next most abundant motif relative to AT/TA was CT/GA, and CGC was the most abundant motif among tri-nucleotide repeats. The di-, tri-, and tetra-nucleotide repeat patterns were different from that of the reference Nipponbare genome (McCouch et al. [2002]; Zhou et al. [2005]), and also differed from that of wheat (Weng et al. [2005]). A total of 90.1% of SSR motifs in the Tongil genome were from *indica*, 9.6% were from*japonica*, and 0.3% were from an unknown genome (Additional file 10: Figure S6).

Distribution of Yield-related Genes in the Tongil genome

One of the most important aims of this study was to explore which regions of the *indica* and*japonica* parental genomes have introgressed into the Tongil variety to provide its high-yield potential. Tongil is morphologically characterized by short plant height, lodging resistance, open plant architecture, medium-long erect leaves, thick leaf sheaths and culms, relatively long panicles, and easily shattered grain (Chung and Heu [1980]) (Figure 1). Although these phenotypic characteristics affect Tongil's high-yield potential, to date we have no molecular genetic evidence regarding the nature of these traits, with the exception of semi-dwarf gene 1 (*sd1*) (Chung and Heu [1980]). Therefore, we analyzed several well-characterized genes associated with high yield potential in the Tongil genome: *sd1* (Nagano et al. [2005]; Sasaki et al. [2002]; Monna et al.[2002]), *Ghd7* (Liu et al. [2013]; Xue et al. [2008]), *Gn1a* (Ashikari et al. [2005]; Miura et al.[2010]), *qSW5* (Yan et al. [2011]; Shomura et al. [2008]), *GS3* (Takano-Kai et al. [2009]; Fan et al.[2006]), and *GW2* (Li et al. [2010]; Song et al. [2007]).

Sd1 (Semi-Dwarf Stature)

Semi-dwarf stature is one of the main genetic contributors to the success of the Green Revolution. The introduction of semi-dwarf genes increased yield by conferring lodging resistance, which enabled greater input of nitrogen fertilizer. Tongil was the first variety into which the *sd1* allele was introduced in South Korea. Analysis of the *sd1* gene, which encodes GA20ox-2 in Tongil and its parents, revealed that Tongil received its *sd1* from an *indica* parent, IR8 or TN1; this allele contained a 383-bp deletion resulting in a frame-shift to form a stop codon (Figure 3A). We also confirmed other *sd1* alleles derived from the native semi-dwarf rice variety Jikkoku (G281T) and the γ-ray–

induced varieties Reimei (C1045G) and Calose76 (C796T) (Monna et al. [2002]; Sasaki et al.[2002]). However, Yukara, the *japonica* parent of Tongil, did not have any *sd1* alleles.

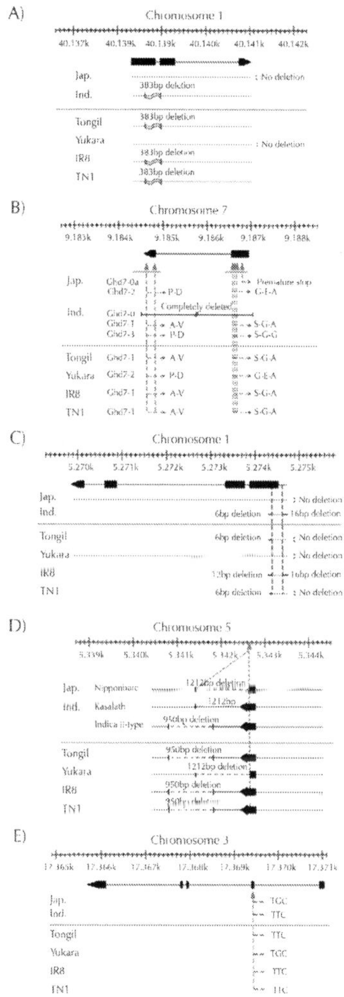

Figure 3: *Indica/japonica*region comparisons of high yield-related alleles or QTLs. **A)***sd1*, **B)***Ghd7*, **C)***Gn1a*, **D)***qSW5*, and **E)***GS3*. Black arrows and box regions represent exons. Vertical, dashed lines refer to the same position in the genome or gene region.

Ghd7 (Grain Number, Plant Height, and Heading Date)

A gene encoding a CCT domain protein, *Ghd7*, is an important regulator of potential yield, plant height, and heading date in rice. Plant height and panicle size are increased under long-day conditions by the delay in heading date resulting from increased *Ghd7* expression. *Ghd7* has five natural variant haplotypes (Xue et al. [2008]). Among these, Tongil possesses the *Ghd7-1* allele (A-G-S-V-A) derived from *indica* parent IR8 or TN1 (Figure 3B), which is considered to be the original, fully-functional wild-type allele; plants with this allele are relatively tall, late heading, have large panicles, and are widely grown. By contrast, the *japonica* parent of Tongil, Yukara, has the *Ghd7-2* allele (A-E-G-D-P), which is weaker than *Ghd7-1* and is found in temperate *japonica* varieties.

Gn1a (Grain Number on Chromosome 1)

Gn1a is one of the most effective QTLs for increasing grain number. It is predicted to encode a cytokinin oxidase/dehydrogenase (OsCKX2). Habataki, an *indica* rice variety, has a 16-bp deletion in the 5′ UTR, a 6-bp deletion in the first exon, and three amino acid substitutions in the first and fourth exons of this gene. In addition, an 11-bp deletion in the third exon has been detected in the high-yielding rice variety 5150 (Ashikari et al. [2005]). Comparisons of DNA sequences between Tongil and parent varieties revealed that the Tongil sequence was identical to the TN1 allele, which had only a 6-bp deletion in the first exon and no 16 bp deletion in the 5′ UTR, as in Habataki. On the other hand, IR8 contained a 16-bp deletion in the 5′ UTR and a 12-bp deletion in the first exon, distinct from the pattern in the TN1 allele. We could not identify any variation in Yukara, which has the same allele sequence as Nipponbare (Figure 3C).

qSW5 (QTL for Seed Width on Chromosome 5)

qSW5 is responsible for seed width; the product of this gene controls cell number in the outer glume of the rice flower. The gene product increases seed width and seed weight by enlarging sink size. The Nipponbare-type allele, which contains a 1,212-bp deletion, is a loss-of-function allele relative to the Kasalath-type allele (Shomura et al. [2008]). In addition, the *indica* II-type allele has a 950-bp deletion relative to the Kasalath allele (Yan et al. [2011]). Comparisons of the*qSW5* alleles between Tongil and parental varieties revealed that Tongil, IR8, and TN1 have the*indica* II-type allele, whereas Yukara has the Nipponbare allele (Figure 3D).

GS3 (Grain Length and Weight; Grain Size 3)

GS3, which encodes a PEPB-like domain protein, was cloned from a QTL for grain length and weight on chromosome 3 in rice. A C-to-A substitution in the second exon of the *GS3* gene is strongly associated with grain length and width: the A-allele confers significantly longer and thinner grains than the C-allele (Takano-Kai et al. [2009]; Fan et al. [2006]). Tongil possesses an A-allele originating from an *indica* parent, IR8 or TN1, whereas the *japonica* parent, Yukara has the C-allele (Figure 3E). In the case of another gene that controls grain width, *GW2*, there were no SNPs detected among any of the strains we sequenced or Nipponbare, indicating that GW2 is a highly conserved gene in rice and even in *Zea mays* (Li et al. [2010]; Song et al. [2007]).

DISCUSSION

In this study, we used high-depth NGS analysis to demonstrate that the Tongil genome is composed of 91.8% *indica*, 7.9% *japonica*, and 0.3% unknown genome. The amounts and types of genes and SSRs in the Tongil genome were very similar to its genomic composition with respect to*indica* or *japonica* origin. This deviation from the expectation that about one-fourth of the Tongil genome originated from the *japonica* parent is likely due to the results of selection during the breeding process and/or to segregation distortion in favor of the *indica* genome because *indica*-type alleles and plants are favored among hybrid progenies from *indica/japonica* crosses (Harushima et al. [1996]; Lin et al. [1992]).

Tongil rice is highly successful in terms of grain yield in South Korea, although Korean climatic environments are not favorable to the cultivation of typical *indica* varieties (Chung and Heu[1991]). This may be attributable to its heightened adaptability compared to most *indica* varieties, perhaps due to the partial incorporation of the *japonica* parental genome.

From an agronomic viewpoint, rice yield is determined by the integration of four yield components: number of panicles per unit area, number of grains per panicle, filled grain ratio, and grain weight. Tongil is a heavy-panicle variety with more grains per panicle than its parental varieties (Chung and Heu [1991]). We manually sequenced the yield-related genes *sd1, Gn1a, Ghd7, GS3, qSW5,* and *GW2* to determine which alleles came from the *indica* and *japonica* parents (Figure 3; Additional file 11: Table S5). The *sd1, Gn1a,* and *Ghd7* alleles of Tongil originated from the *indica* parents, as did the *GS3, qSW5,* and *GW2* alleles of Tongil, all of which are involved in determining seed size. Although Tongil and its *indica* parents share the same allele for these three genes, the seed shape of Tongil is closer to TN1 than that of IR8 (Figure 1). Thus, another gene or epistatic interaction may be involved in determining seed shape in the Tongil cultivar (Yan et al. [2011]). In fact, all of the alleles identified in Tongil were the same as those in TN1, one of the *indica* parents. Therefore, it is unlikely that a greater understanding of the high yield potential of Tongil could be achieved by analyzing these six yield-related genes. Complex genetic systems, including unknown genes and epistatic interactions, should be investigated in future studies.

Since the initial success of Tongil rice in Korea, numerous Tongil-type varieties of similar parentage or that were bred using Tongil rice as one of the parents have been developed to address future needs for food security. We predict that genomic information, including the SNP data provided in this study, will facilitate the efficient breeding of these and other Tongil-type varieties.

CONCLUSIONS

We determined the genome structure of Tongil rice, a successful cultivar derived from *indica* × *japonica* hybridization in Korea. Analyses of genome composition and genetic factors of Tongil rice demonstrate that the Tongil genome is derived mostly from the *indica* genome, with a small portion of *japonica* genome introgression. The approach used in this study to determine the parental origins of specific genome segments is applicable to the genomic dissection of agricultural breeding lines or varieties of diverse parental origins.

METHODS

Plant Materials

Plant lines subjected to whole-genome resequencing in the present study included Tongil (SNU accession no. 260697) and its parental lines: Yukara, an early maturing temperate *japonica* cultivar (RDA-Genebank Information Center accession no. IT004665); Taichung native 1 (TN1), the first semi-dwarf *indica* variety with high adaptability (RDA-Genebank Information Center accession no. IT004120); and IR8, an improved high-yielding semi-dwarf variety developed at the International Rice Research Institute (IRRI, IRTP 195). The Tongil variety was developed through a three-way cross, IR8//Yukara/TN1. With generation advancement after the cross, the most promising line, IR667-98-1-2, was selected and released to farmers in Korea under the name 'Tongil' (Chung and Heu [1991]).

Whole-Genome DNA Sequencing

Four rice varieties were sequenced: Tongil and its parental varieties, Yukara, IR8, and TN1. Whole-genome shotgun sequencing of the four rice genomes was performed using the Illumina/Solexa GAII system. DNA sequencing, including construction of shotgun DNA libraries, was performed according to the methods recommended by the manufacturer (Illumina, San Diego, CA, USA). Briefly, whole-genome DNA shotgun paired-end sequencing libraries were generated by fragmentation of DNA into 500-bp segments using a Covaris DNA shearing machine (Covaris, CA, US), followed by ligation of paired-end adapters ligation of 53 and 68 bp for sequencing on the FlowCell, size selection of the adapter-ligated fragments within the desired size range (500–600 bp), and PCR enrichment using complete primer constructs required for binding and clustering on the FlowCell. Illumina GAII sequencing was performed by identifying the emission color of single-base extensions on the FlowCell.

ABBREVIATIONS

SNP: Single nucleotide polymorphism

NGS: Next generation sequencing

IRGSP: International rice genome sequencing project

SSR: Simple sequence repeat

GO: Gene ontology

SEG-map: Sequencing enabled genotyping for mapping recombination populations

Competing interests

The authors have no potential competing interests to declare.

Authors' contributions

BK and HK conceived of the study and participated in its design. IC and BC performed bioinformatic analysis and data processing. BK and JL collected samples and phenotype data. DK, BK, GL, and JS analyzed the data and helped to draft the manuscript. TY, KK, DK, and JC helped to revise the manuscript. All authors read and approved the final manuscript.

ACKNOWLEDGMENTS

This work was supported by a grant from the Next-Generation BioGreen 21 Program (Plant Molecular Breeding Center No. PJ008125), Rural Development Administration, Republic of Korea.

ACCESSION CODES

Raw sequence data obtained in our study have been submitted to the NCBI Short Read Archive (Takano-Kai et al.) with the following accession numbers: Tongil [SRA: SRR923809, SRA: SRR923810], IR8 [SRA: SRR921498], TN1 [SRA: SRR921505], and Yukara [SRA:

SRR925387].

REFERENCES

1. Ashikari M, Sakakibara H, Lin S, Yamamoto T, Takashi T, Nishimura A, Angeles ER, Qian Q, Kitano H, Matsuoka M (2005) Cytokinin oxidase regulates rice grain production. Science 309(5735):741-745 doi: 10.1126/science.1113373
2. Chen H, Xie W, He H, Yu H, Chen W, Li J, Yu R, Yao Y, Zhang W, He Y, Tang X, Zhou F, Deng XW, Zhang Q (2014) A high-density SNP genotyping array for rice biology and molecular breeding. Mol Plant 7(3):541-553 doi:10.1093/mp/sst135
3. Cheng SH, Zhuang JY, Fan YY, Du JH, Cao LY (2007) Progress in research and development on hybrid rice: a super-domesticate in China. Ann Bot 100(5):959-966 doi:10.1093/aob/mcm121
4. Chung GS, Heu MH (1980) Status of japonica-indica hybridization in Korea. In: Selected papers from the 1979 International Rice Research. International Rice Research Institute, Manila. pp 135-152
5. Chung GS, Heu MH (1991) Improvememt of Tongil-Type Rice Cultivars from Indica Japonica Hybridization in Korea. In: Biotechnology in Agriculture and Forestry 14. Springer-Verlag Berlin Heidelberg, Germany. pp 105-112
6. Conesa A, Gotz S, Garcia-Gomez JM, Terol J, Talon M, Robles M (2005) Blast2GO: a universal tool for annotation, visualization and analysis in functional genomics research. Bioinformatics 21(18):3674-3676 doi:10.1093/bioinformatics/bti610
7. Fan C, Xing Y, Mao H, Lu T, Han B, Xu C, Li X, Zhang Q (2006) GS3, a major QTL for grain length and weight and minor QTL for grain width and thickness in rice, encodes a putative transmembrane protein. Theor Appl Genet 112(6):1164-1171
8. doi:10.1007/s00122-006-0218-1
9. Feltus FA, Wan J, Schulze SR, Estill JC, Jiang N, Paterson AH (2004) An SNP resource for rice genetics and breeding based on subspecies indica and japonica genome alignments. Genome Res 14(9):1812-1819 doi:10.1101/gr.2479404

10. Harushima Y, Kurata N, Yano M, Nagamura Y, Sasaki T, Minobe Y, Nakagahra M (1996) Detection of segregation distortions in an indica-japonica rice cross using a high-resolution molecular map. Theor Appl Genet 92(2):145-150 doi:10.1007/BF00223368
11. Huang X, Feng Q, Qian Q, Zhao Q, Wang L, Wang A, Guan J, Fan D, Weng Q, Huang T, Dong G, Sang T, Han B (2009) High-throughput genotyping by whole-genome resequencing. Genome Res 19(6):1068-1076 doi:10.1101/gr.089516.108
12. Huang X, Kurata N, Wei X, Wang ZX, Wang A, Zhao Q, Zhao Y, Liu K, Lu H, Li W, Guo Y, Lu Y, Zhou C, Fan D, Weng Q, Zhu C, Huang T, Zhang L, Wang Y, Feng L, Furuumi H, Kubo T, Miyabayashi T, Yuan X, Xu Q, Dong G, Zhan Q, Li C, Fujiyama A, Toyoda A, Lu T, Feng Q, Qian Q, Li J, Han B (2012) A map of rice genome variation reveals the origin of cultivated rice. Nature 490(7421):497-501 doi:10.1038/nature11532
13. Initiative TAG (2000) Analysis of the genome sequence of the flowering plant Arabidopsis thaliana. Nature 408(6814):796-815
14. Khush GS (1999) Green revolution: preparing for the 21st century. Genome 42(4):646-655
15. Li Q, Li L, Yang X, Warburton ML, Bai G, Dai J, Li J, Yan J (2010) Relationship, evolutionary fate and function of two maize co-orthologs of rice GW2 associated with kernel size and weight. BMC Plant Biol 10:143 doi:10.1186/1471-2229-10-143
16. Lin SY, Ikehashi H, Yanagihara S, Kawashima A (1992) Segregation distortion via male gametes in hybrids between Indica and Japonica or wide-compatibility varieties of rice (Oryza sativa L). Theoret Appl Genetics 84(7–8):812-818 doi:10.1007/BF00227389
17. Liu T, Liu H, Zhang H, Xing Y (2013) Validation and characterization of Ghd7.1, a major QTL with pleiotropic effects on spikelets per panicle, plant height, and heading date in rice (Oryza sativa L.). J Integr Plant Biol, doi:10.1111/jipb.12070
18. Lu BR, Zheng KL, Qian HR, Zhuang JY (2002) Genetic differentiation of wild relatives of rice as assessed by RFLP analysis. Theor Appl Genet 106(1):101-106 doi:10.1007/s00122-002-1013-2
19. McCouch SR, Teytelman L, Xu Y, Lobos KB, Clare K, Walton M, Fu B, Maghirang R, Li Z, Xing Y, Zhang Q, Kono I, Yano M,

Fjellstrom R, DeClerck G, Schneider D, Cartinhour S, Ware D, Stein L (2002) Development and mapping of 2240 new SSR markers for rice (Oryza sativa L.). DNA Res 9(6):199-207

20. McNally KL, Childs KL, Bohnert R, Davidson RM, Zhao K, Ulat VJ, Zeller G, Clark RM, Hoen DR, Bureau TE, Stokowski R, Ballinger DG, Frazer KA, Cox DR, Padhukasahasram B, Bustamante CD, Weigel D, Mackill DJ, Bruskiewich RM, Ratsch G, Buell CR, Leung H, Leach JE (2009) Genomewide SNP variation reveals relationships among landraces and modern varieties of rice. Proc Natl Acad Sci U S A 106(30):12273-12278doi:10.1073/pnas.0900992106
21. Miura K, Ikeda M, Matsubara A, Song XJ, Ito M, Asano K, Matsuoka M, Kitano H, Ashikari M (2010) OsSPL14 promotes panicle branching and higher grain productivity in rice. Nat Genet 42(6):545-549 doi:10.1038/ng.592
22. Monna L, Kitazawa N, Yoshino R, Suzuki J, Masuda H, Maehara Y, Tanji M, Sato M, Nasu S, Minobe Y (2002) Positional cloning of rice semidwarfing gene, sd-1: rice "green revolution gene" encodes a mutant enzyme involved in gibberellin synthesis. DNA Res 9(1):11-17
23. Morishima H, Oka HI (1981) Phylogenetic Differentiation of Cultivated Rice.22. Numerical Evaluation of the Indica-Japonica Differentiation. Jpn J Breed 31(4):402-413
24. Nagano H, Onishi K, Ogasawara M, Horiuchi Y, Sano Y (2005) Genealogy of the "Green Revolution" gene in rice. Genes Genet Syst 80(5):351-356
25. (2009) Climate change: impact on agriculture and costs of adaptation. Food policy report, International Food Policy Research Institute, Washington, D.C.
26. Sano R, Morishima H (1992) Indica-Japonica Differentiation of Rice Cultivars Viewed from Variations in Key Characters and Isozymes, with Special Reference to Landraces from the Himalayan Hilly Areas. Theor Appl Genet 84(3–4):266-274
27. Sasaki A, Ashikari M, Ueguchi-Tanaka M, Itoh H, Nishimura A, Swapan D, Ishiyama K, Saito T, Kobayashi M, Khush GS, Kitano H, Matsuoka M (2002) Green revolution: a mutant gibberellin-synthesis gene in rice. Nature 416(6882):701-702 doi:10.1038/416701a

28. Shen Y, Wan Z, Coarfa C, Drabek R, Chen L, Ostrowski EA, Liu Y, Weinstock GM, Wheeler DA, Gibbs RA, Yu F (2010) A SNP discovery method to assess variant allele probability from next-generation resequencing data. Genome Res 20(2):273-280 doi:10.1101/gr.096388.109
29. Shomura A, Izawa T, Ebana K, Ebitani T, Kanegae H, Konishi S, Yano M (2008) Deletion in a gene associated with grain size increased yields during rice domestication. Nat Genet 40(8):1023-1028 doi:10.1038/ng.169
30. Song XJ, Huang W, Shi M, Zhu MZ, Lin HX (2007) A QTL for rice grain width and weight encodes a previously unknown RING-type E3 ubiquitin ligase. Nat Genet 39(5):623-630 doi: 10.1038/ng2014
31. Takano-Kai N, Jiang H, Kubo T, Sweeney M, Matsumoto T, Kanamori H, Padhukasahasram B, Bustamante C, Yoshimura A, Doi K, McCouch S (2009) Evolutionary history of GS3, a gene conferring grain length in rice. Genetics 182(4):1323-1334 doi:10.1534/genetics.109.103002
32. Weng DX, Xu SC, Lin RM, Wan AM, Li JP, Wu LR (2005) Microsatellite marker linked with stripe rust resistant gene Yr9 in wheat. Yi Chuan Xue Bao 32(9):937-941
33. Xue W, Xing Y, Weng X, Zhao Y, Tang W, Wang L, Zhou H, Yu S, Xu C, Li X, Zhang Q (2008) Natural variation in Ghd7 is an important regulator of heading date and yield potential in rice. Nat Genet 40(6):761-767 doi:10.1038/ng.143
34. Yan S, Zou G, Li S, Wang H, Liu H, Zhai G, Guo P, Song H, Yan C, Tao Y (2011) Seed size is determined by the combinations of the genes controlling different seed characteristics in rice. Theor Appl Genet 123(7):1173-1181 doi:10.1007/s00122-011-1657-x
35. Zhao K, Wright M, Kimball J, Eizenga G, McClung A, Kovach M, Tyagi W, Ali ML, Tung CW, Reynolds A, Bustamante CD, McCouch SR (2010) Genomic diversity and introgression in O. sativa reveal the impact of domestication and breeding on the rice genome. PLoS One 5(5):e10780
36. Zhao QA, Huang XH, Lin ZX, Han B (2010b) SEG-Map: A Novel Software for Genotype Calling and Genetic Map Construction from Next-generation Sequencing. Rice 3(2–3):98-102 doi:10.1007/s12284-010-9051-x

37. Zhou LJ, Ao GH, Wu XJ, Li SG (2005) SSR markers linked with early stability in rice. Yi Chuan Xue Bao 32(8):837-845

Chapter 8

Unraveling the Secrets of Rice Wild Species

Ehsan Shakiba[2] and Georgia C. Eizenga[1]

[1] USDA-ARS Dale Bumpers National Rice Research Center, Stuttgart, AR, USA

[2] University of Arkansas Rice Research and Extension Center, Stuttgart, AR, USA

INTRODUCTION

The world is facing a new challenge with global population predicted to plateau at nine billion people by the middle of this century (Godfray et al. 2010). Increasing food production to feed the world's population is an even greater challenge considering that agriculture is experiencing greater competition for land, water and energy, as well as, the effects of substantial climate change and the unintended effects of crop production on the environment. Part of the solution to increasing food production on the same or less cultivated land lies in exploiting the subset of genes lost during the domestication process and subsequent targeted breeding. Currently, these valuable genes are found only in the progenitor species genepool for crop cultivars. Cultivated plants having desirable genes were utilized in intensive breeding projects focused on increasing yield for particular environments and management systems but this process has narrowed the genetic diversity (Rausher 2001). For cultivated plants, this unexploited genetic material includes both landraces and the more exotic wild relatives. Improving our understanding of this tertiary gene pool and exploiting it for crop improvement is paramount to meeting the challenges of feeding the world in this century through the integration of classical genetics and genomics-enabled research paradigms.

The loss of genetic diversity can be more problematic for self-pollinated plant species where the rate of cross pollination is below five percent, thus making it more difficult to reintroduce the lost diversity. In the case of the two major grain crops, rice (*Oryza sativa* L.) and wheat (*Triticum aestivum* L.), both self-pollinated, the re-introduction genetic diversity from the wild is central to the continued success of breeding, given that viruses, fungi, and bacteria, three main causal agents of biotic stress, are constantly evolving to cause the breakdown of the host plant's defense mechanisms (Rausher 2001).

Abiotic stress, including salinity, aluminum toxicity and acid sulfate soils, as well as, temperature and drought, complicate the difficulty of improving crop yields, especially in the face of global warming (Brooker 2006; Tilman and Lehman 2001), which makes modern cultivars even more vulnerable. Genetic sources of resistance or tolerance offer the most promising mechanism to protect plants against these unfavorable conditions. Often wild species are not included as parental lines in cultivar development because it is relatively difficult to harness desirable genes by genetic recombination and many undesirable genes are introgressed from the wild parent resulting in inferior yield, undesirable plant architecture, and/or poor grain quality (Tanksley and McCouch

1997). Recent studies, however, in rice (McCouch et al. 2007) and tomato, *Lycopersicon esculentum* Mill. (Grandillo and Tanksley 2003), have shown that wild species contain genomic components that could result in genetic gains in terms of agronomic performance.

The rapid advancement in molecular technologies allows for genotyping plants much more quickly and inexpensively than ever before. The availability of high resolution genotypic information creates the opportunity to further explore an expanding number of accessions in a greater depth, and harness this information to enhance the efficiency and accuracy of introgression. These developments create opportunities not previously possible, to identify molecular markers associated with desirable traits in wild species and transfer these traits into elite lines and/or varieties, as well as, to unravel multi-genic traits for crop improvement (Tanksley and McCouch 1997; McCouch et al. 2012).

Our main objective is to summarize efforts over the past 15 years to identify useful novel alleles in the*Oryza* species that were lost during evolution and domestication, genetically dissect the traits encoded by these alleles through chromosome mapping, and incorporate these traits or alleles into an agronomically useful genetic background. To do this we will (a) briefly describe the relationships among the species in the genus *Oryza*, (b) describe the types of populations that have been developed for mapping desirable traits identified in the wild *Oryza* species to a chromosome location, and (c) summarize the quantitative trait locus (QTL) studies focused on mapping the useful traits and novel alleles to specific locations in the genomes of *Oryza* species.

PHYLOGENY OF THE ORYZA GENUS

The *Oryza* genus includes two cultivated species, Asian rice, *O. sativa*, which is grown throughout the tropical and temperate climates of the world, and African rice, *O. glaberrima*, which is found in sub-Saharan Africa along the Niger River. The 22 wild species composing the *Oryza* genus are characterized by eleven different genomes identified as the A-, B-, C-, D-, E-, F-, G-, H-, J-, K-and L-genomes and arranged in the following 10 genome types AA, BB, CC, BBCC, CCDD, EE, FF, GG, HHJJ and KKLL. Four of the wild *Oryza* species are tetraploid and the remaining 18 are diploid, as well as, the two cultivated species (Table 1).

TABLE 1: Taxonomic classification of *Oryza* species including the chromosome number, genome designation, genome size and distribution for each species

Species†	No. of chromosomes (2n)†	Genome†	Genome size (Mbp)‡	Distribution†
Section Oryza				
Oryza sativa complex				
O. sativa L. (ssp. *japonica*, ssp. *indica*)	24	AA	420, 466	Worldwide
O. nivara Sharma et Shastry	24	AA	448	Tropical and subtropical Asia
O. rufipogon Griff.	24	AA	439, 450	Tropical and subtropical Asia, Tropical Australia
O. glaberrima Steud.	24	$A^{g}A^{g}$	354	West Africa
O. barthii A. Chev.§	24	$A^{g}A^{g}$	411	Africa
O. glumaepatula Steud.	24	$A^{gp}A^{gp}$	464	South and Central America
O. longistaminata A. Chev. et Roehr.	24	$A^{l}A^{l}$	352	Africa
O. meridionalis Ng	24	$A^{m}A^{m}$	435	Tropical Australia
Oryza officinalis complex				
O. punctata Kotschy ex Steud.	24 48	BB, BBCC	423 (BB)	Africa
O. minuta J.S. Presl. ex C.B. Presl.	48	BBCC	1124	Philippines, Papua New Guinea
O. eichingeri A. Peter	24	CC		South Asia and East Africa
O. officinalis Wall ex Watt	24	CC	653	Tropical and subtropical Asia, Tropical Australia
O. rhizomatis Vaughan	24	CC		Sri Lanka

	O. alta Swallen	48	CCDD	1124	South and Central America
	O. grandiglumis (Doell) Prod.	48	CCDD		South and Central America
	O. latifolia Desv.	48	CCDD		South and Central America
	O. australiensis Domin.	24	EE	960	Tropical Australia
Section Brachyantha					
	O. brachyantha A. Chev. et Roehr.	24	FF	338	Central Africa
Section Padia					
Oryza granulata complex					
	O. granulata Nees et Arn. ex Watt	24	GG	862	South and Southeast Asia
	O. meyeriana (Zoll. et (Mor. ex Steud.) Baill.)	24	GG		Southeast Asia
Oryza ridleyi complex					
	O. longiglumis Jansen	48	HHJJ		Irian Jaya, Indonesia, Papua New Guinea
	O. ridleyi Hook. F.	48	HHJJ	1283	South Asia
Oryza schlechteria complex					
	O. coarctata Tateoka	48	KKLL	771	Asian coastal area
	O. schlerchteri Pilger	48	KKLL		Papua New Guinea

[i] - [I] Classification for species, genome designation and distribution based on Brar and Singh (2011), Lu et al. (2014), Sanchez et al. (2013) and Vaughan (2003). The superscripts for the A-genome indicate a variation of the type of A-genome.

[ii] - [‡] Genome size based on the following: *O. sativa* subsp. *japonica* (Goff et al. 2002), *O. sativa*subsp. *indica* (Yu et al. 2002) and *Oryza* species (Ammiraju et al. 2010).

[iii] - [§] *O. barthii* is also classified as *O. breviligulata* A. Chev. et Roehr.

Rice is the only major cereal found in the ancient lineage of the Bambusoideae and is currently placed in the subfamily Erhartoideae. Historically, the grass family, Poaceae, is thought to have evolved about 70-55 mya (million years ago) with the tribes Oryzeae and Pooideae (wheat and oats) diverging about 35 mya [reviewed by Kellogg (2009) and Vaughan et al. (2008)]. The Oryzinae and Zizaninae subtribes diverged about 20-22 mya and the *Oryza* and *Leersia* genera about 14.2 mya. The genus *Oryza* is divided into three sections: *Padia, Brachyantha* and *Oryza*. *Padia* includes the forest-dwelling *Oryza,* which are distributed into the *O. granulata* (GG), *O. ridleyi* (HHJJ) and *O. schlechteria* (KKLL) complexes. The *O. granulata* complex is thought to have diverged from the other *Oryza* species about 8 mya. *O. brachyantha* (FF) is the only species in the section *Brachyantha*. This species is widely distributed across Africa, growing in iron-pan rock pools.

Section *Oryza* consists of two species complexes, the *O. officinalis* complex with the B-, C-, D-and E-genomes and the *O. sativa* complex, which includes all the A-genome species. Within the *O. officinalis* complex, *O. australiensis* (EE) is the most diverged and *O. eichingeri* (CC) appears to be the most basal of the C-genome species.

The species in the *O. sativa* complex prefer full sun, and grow near lakes, rivers and seasonal pools of water. Molecular data suggests that *O. meridionalis* diverged from the other A-genome species about 2 mya. Also, the perennial African species, *O. longistaminata,* diverged from the Asian A-genome species about the same time period, 2-3 mya. The second divergence between the Asian and African A-genome species, *O. barthii* and *O. glaberrima,* occurred 0.6 to 0.7 mya. More recently, possibly about 0.4 mya (or more than 0.2 mya), the *O. rufipogon* clade(s) that eventually diverged into the *O. sativa* subspecies (subsp.) *Japonica* and *Indica*. Later, the *Indica* subspecies differentiated into the*indica* and *aus* subpopulations and the *Japonica* subspecies into the *aromatic* (Group V), *tropical japonica* and *temperate japonica* subpopulations (Garris et al. 2005, Zhao et al. 2011, Huang et al. 2012). Archaeobotanical evidence from spikelet bases and changes in grain size document this domestication process (Fuller et al. 2010). Recently, based on genome sequences of 446 geographically diverse *O. rufipogon* accessions, Huang et al. (2012) further subdivided *O. rufipogon*accessions into three major *O. rufipogon* clades: one closely aligned with *O. sativa* subsp. *japonica,* one aligned with *O. sativa* subsp. *indica,* and the third clade was independent of *O. sativa*. Furthermore, as part of this study, a neighbor-joining tree constructed from sequence differences of 15 representative A-genome accessions suggested within *Indica,* different *O. rufipogon* clades were associated with the *aus* and *indica* subpopulations, whereas the three *Japonica* subpopulations arose from a single *O. rufipogon* clade. This phylogenetic tree also supported the aforementioned genetic distance between *O. meridionalis*, *O. longistaminata*, *O. barthii* and *O. glaberrima*.

Rice, *O. sativa*, the first monocot plant with a reference genome, is the central comparative genomics model for all grasses, and has been compared to all major cereals. To lay the foundation for interrogating the rice wild relatives, 18 bacterial artificial chromosome (BAC) libraries for 16 different *Oryza* species spanning all 10 *Oryza* genome types including the AA-genome species (*O. nivara*, *O. rufipogon*, *O. glaberrima*, *O. barthii*, *O. glumaepatula*, *O. longistaminata*, *O. meridionalis*), *O. punctata* (BB), *O. officinalis* (CC), *O. minuta* (BBCC), *O. alta* (CCDD), *O. australiensis* (EE), *O. brachyantha* (FF), *O. granulata* (GG), *O. ridleyi* (HHJJ) and *O. coarctata* (HHKK), were generated through the *Oryza* Map Alignment Project (OMAP) as summarized by Ammiraju et al. (2010). Subsequently, the International OMAP consortium was formed in 2007 to (a) generate reference sequences and transcriptome data sets of the eight A-genome species and representative species of the other genome types, (b) generate advanced mapping populations for the A-genome species, and (c) identify naturally occurring populations of the wild *Oryza* species for diversity and evolutionary analyses, as well as, conservation (Jacquemin et al. 2013; Sanchez et al. 2013). The species included in the sequencing effort were A-genome species (*O. nivara*, *O. rufipogon*, *O. barthii*, *O. glaberrima*, *O. glumaepatula*, *O. longistaminata*, *O. meridionalis* and both *O. sativa* subsp. *indica* and subsp. *japonica*), *O. punctata* (BB), C-genome species (*O. officinalis*, *O. eichingeri*, *O. rhizomatis*), *O. australiensis* (EE), *O. brachyantha* (FF), *O. granulata* (GG), and the outgroup, *Leersia perrieri*. To date, the sequencing of nine genomes and *L. perrieri* has been completed, in addition to the established reference sequences for *O. sativa* subsp. *japonica* and subsp. *indica* genomes (Wing 2013). Currently, two additional *O. sativa* subsp. *indica* cultivars are being sequenced representing the *aus* (DJ123) and *indica* (IR64) subpopulations (McCombie 2013).

Methods for Developing Oryza Interspecific Mapping Populations

Traits are classified as either qualitative or quantitative traits. Qualitative traits are controlled by one or a few genes with major effects while quantitative characters are controlled by many genes with minor effects (Poehlman and Sleper 1994). Identification of genes associated with quantitative traits is always more complicated compared to those involving qualitative traits.

Interspecific and intergenomic hybridization, hybridization between species with the same or different genomes, have been used to transfer desirable genes or QTL associated with simple or complex traits from wild species into a cultivated genetic background (Brar and Khush 1997; Dalmacio et al. 1995;Tanksley and McCouch 1997). Nevertheless hybridization success can be hindered by genomic incompatibilities and sterility barriers (Ishii et al. 1994; McCouch et al. 2007; Wang et al. 2005). The utilization of embryo rescue and other methods of producing viable and fertile hybrids combined with robust molecular markers and associated computational and statistical analyses, led to the successful generation of interspecific genetic populations that were used to link desirable traits to molecular markers and subsequent identification of the actual genes controlling the traits of interest (Ali et al. 2010; Chen et al. 2010; Ghesquière et al. 1997; Guo et al. 2013; Lexer and Fay 2005; McCouch et al. 2007). Six types of mapping populations are generated from interspecific crosses between *Oryza*species and *O. sativa* including (a) recombinant inbred line (RIL), (b) advanced backcross (AB), (c) backcross inbred line (BIL), (d) chromosome segment substitution line (CSSL), (e) near isogenic line (NIL) and (f) multi-parent advanced generation inter-cross (MAGIC). A discussion of each of these populations follows and examples are included in the third section describing agronomically important traits attributed to the *Oryza* species donor.

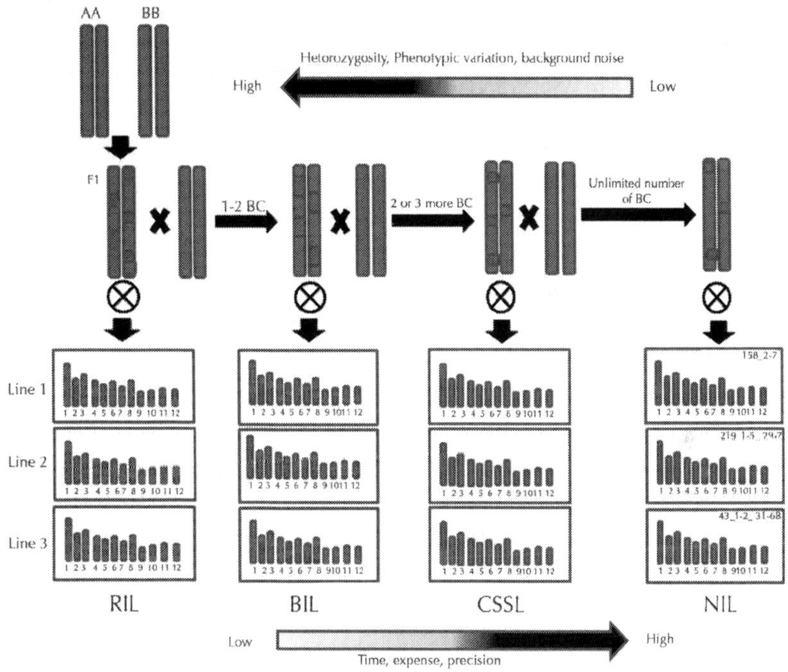

Figure 1: A comparison of the methods for creating primary and advanced bi-parental mapping populations, including recombinant inbred lines (RILs), backcross inbred lines (BILs), chromosome segment substitution lines (CSSLs) and near isogenic lines (NILs) as summarized by Fukuoka et al. (2010). Also shown are the number of backcrosses (BC) required and the genotypes of the lines obtained by each method. Karyotypes of the three CSSLs illustrate how chromosome 1 of the donor can be introgressed into the recurrent parent. The three NIL genotypes are based on the JeffersonNILs, each with a different *O. rufipogon* (IRGC105491) introgression selected for a different yield QTL (Imai et al. 2013).

Recombinant Inbred Line (RIL) Population

RIL populations have been the most common type of mapping population used in rice genetics and breeding when both parents are *O. sativa* but a limited number of interspecific populations have been reported. To develop a RIL population, two contrasting cultivars or accessions for the trait(s) to be mapped are crossed together to

permanent mapping populations for precise QTL mapping, and for direct or indirect use in variety development, especially when the parents used to develop the population are the source of agronomically useful traits (Bandillo et al. 2013). MAGIC populations are developed by systematically crossing several F_1 hybrids involving four to sixteen different parental lines to create a set of double crosses, then systematically crossing the double cross hybrids to create a set of 4-, 8-or 16-way crosses. As the final step, the lines composing the population are advanced four or more generations by single seed descent to obtain a set of advanced intercrossed lines (AILs).Bandillo et al. (2013) reported four different types of MAGIC populations being developed in rice (*O. sativa*) at the International Rice Research Institute (IRRI) which are described as (1) *Indica* MAGIC composed of 1,831 S_8 AILs; (2) MAGIC Plus with 2,214 S_6 AILs; (3) *Japonica* MAGIC with approximately 400 S_6 AILs; and (4) MAGIC Global with 1,402 AILs in the S_5 generation. Currently, a Wild MAGIC population is being developed by a team at IRRI (K. Jena, H. Leung, K. McNally) in collaboration with J. Hibberd (University of Cambridge, U.K.), and I. Mackay (NIAB, Cambridge, U.K.) using multiple accessions of all eight A-genome species (McNally, personnel communication). In most cases, for this population, the initial crosses had *O. sativa* as the female parent, and the goal is to produce 16-way crosses with highly mixed genomes.

Useful Agronomic Traits Mapped in Oryza Species and Transferred into Cultivated Rice

Table 2: Summary of QTLs for improved yield and yield components attributed to the non-*O. sativa* parent

Trait	Donor species	Donor accession	Recurrent parent(s)	Type of mapping population†	QTL mapping analysis‡	Chromosome location§	Type of marker¶	Reference
Vegetative Growth Stages								
Days to flowering	*O. australiensis*	IRGC100882	IR31917-45-3-2	IL		10	RFLP	Ishii et al. (1994)
Days to heading	*O. glaberrima*		IR64	BC$_{2F}$3	SPA	2, 10	SSR, STS	Bimpong et al. (2011)
	O. glaberrima	IRGC103544	Milyang 23	BC$_{3F}$2	SPA	1, 4, 7, 8	SSR	Suh et al. (2005)
	O. glumaepatula		Taichung 65	BC$_{4F}$2		7	RFLP	Sanchez et al. (2003)
	O. grandiglumis	IRGC101[illegible]54	Hwaseongbyeo	AB-QTL	SPA	6	SSR	Yoon et al. (2006)
	O. meyeriana	Y73	IR24	RIL	CIM	6, 7, 8, 11	SSR, STS	Chen et al. (2012)
	O. nivara	IRGC100398	Bengal	AB-QTL	MIM	3, 6	SSR	Eizenga et al. (2013)
	O. nivara	IRGC100398	Bengal	AB-QTL	MIM	3, 4, 6, 8	SSR	Eizenga et al. (2013)
	O. nivara	IRGC100195	M-202	AB-QTL	MIM	3, 8	SSR	Eizenga et al. (accepted)

	O. rufipogon	IRGC105491	Jefferson	AB-QTL	SPA, IM, CIM	1, 2, 3, 4, 10	RFLP, SSR	Thomson et al. (2003)
	O. rufipogon	IRGC105491	IR64	AB-QTL	SPA, IM, CIM	2, 7	RFLP, SSR	Septiningsih et al. (2003)
	O. rufipogon	IRGC105491	V20A, V20B	B_C2	ANOVA	6, 12	RFLP	Xiao et al. (1998)
	O. rufipogon	IRGC105491	Hwaseongbyeo	AB-QTL, NIL	SPA, IM, ANOVA	6, 9	SSR	Jin et al. (2009), Xie et al. (2008)
	O. rufipogon	IRGC105491	MR219	AB-QTL	CIM	3	SSR	Wickneswari et al. (2012)
	O. rufipogon	W1944	Hwayeongbyeo	IL	SPA, IM	1	SPA, IM	Yuan et al.(2009)
Days to maturity	*O. rufipogon*	IRGC105491	IR64	AB-QTL	SPA, IM, CIM	4, 7, 8	RFLP, SSR	Septiningsih et al. (2003)
	O. rufipogon	IRGC105491	V20A, V20B	B_C2	ANOVA	6, 12	RFLP	Xiao et al. (1998)
	O. rufipogon	IRGC105491	MR219	AB-QTL	CIM	4, 6	SSR	Wickneswari et al. (2012)
Seedling height	*O. rufipogon*	W1944	Hwayeongbyeo	RIL	SPA, CIM	1	SSR	Lee et al. (2005)
Culm length	*O. glaberrima*	IRGC103544	Milyang 23	$BC_{3F}2$	SPA	2, 10	SSR	Suh et al. (2005)
	O. grandiglumis	IRGC101154	Hwaseongbyeo	AB-QTL	SPA	1, 4	SSR	Yoon et al. (2006)
	O. minuta	IR71033-121-15	Junambyeo	$_{F2:}3$	SPA, IM	6, 7, 12	SSR, STS	Rahman et al. (2007)
	O. rufipogon	W1944	Hwayeongbyeo	RIL	SPA, CIM	1, 6	SSR	Lee et al. (2005)
	O. rufipogon	IRGC105491	MR219	AB-QTL	CIM	1, 3, 9	SSR	Wickneswari et al. (2012)
	O. rufipogon	W1944	Hwayeongbyeo	IL	SPA, IM,	1, 12	SPA, IM	Yuan et al.(2009)
Plant height	*O. glaberrima*		IR64	$BC_{2F}3$	SPA	1	SSR, STS	Bimpong et al. (2011)

	O. longistaminata		RD23	$BC_{7F}2$	CIM	1	SSR	Chen et al. (2009)
	O. nivara	IRGC100898	Bengal	AB-QTL	MIM	1	SSR	Eizenga et al. (2013)
	O. nivara	IRGC104705	Bengal	AB-QTL	MIM	1, 12	SSR	Eizenga et al. (2013)
	O. rufipogon	IRGC105491	Jefferson	AB-QTL	SPA, IM, CIM	1	RFLP, SSR	Thomson et al. (2003)
	O. rufipogon	IRGC105491	IR64	AB-QTL	SPA, IM, CIM	1, 4, 10, 11	RFLP, SSR	Septiningsih et al. (2003)
	O. rufipogon	IRGC105491	Hwaseongbyeo	AB-QTL, NIL	SPA, IM, ANOVA	7, 9	SSR	Jin et al. (2009), Xie et al. (2008)
	O. rufipogon	IRGC105491	MR219	AB-QTL	CIM	1, 3, 9	SSR	Wickneswari et al. (2012)
	O. rufipogon	IC22015	IR 58025A	AB-QTL	IM, CIM	1	SSR	Marri et al. (2005)
Plant type (Culm habit or tiller angle)	*O. nivara*	IRGC100898	Bengal	AB-QTL	MIM	9	SSR	Eizenga et al. (2013)
	O. nivara	IRGC104705	Bengal	AB-QTL	MIM	9	SSR	Eizenga et al. (2013)
	O. nivara	IRGC100195	M-202	AB-QTL	MIM	9	SSR	Eizenga et al. (accepted)
Flag leaf length	*O. minuta*	IR71033-121-15	Junambyeo	$F_{2:3}$	SPA, IM	8, 9	SSR, STS	Rahman et al. (2007)
Third node width	*O. rufipogon*	W1944	Hwayeongbyeo	RIL	SPA, CIM	1	SSR	Lee et al. (2005)
Tiller number	*O. glaberrima*		IR64	$BC_{2F}3$	SPA	2, 7	SSR, STS	Bimpong et al. (2011)
	O. glumaepatula	RS-16	BG90-2	$BC_{2F}2$	SPA, IM	4, 5, 7, 8, 11	SSR, STS	Brondani et al. (2002)
	O. glumaepatula	RS-16	Cica8	$BC_{2F2:}9$	CIM	7, 11	SSR	Rangel et al. (2013)
	O. minuta	IR71033-121-15	Junambyeo	$F_{2:3}$	SPA, IM	3	SSR, STS	Rahman et al. (2007)

Hybrid breakdown locus	*O. nivara*	IRGC105444	Koshihikari	$BC_{4F}3$		2	SSR, SNP	Miura et al. (2008)
Panicle fertility	*O. glaberrima*		IR64	$BC_{2F}3$	SPA	2, 10	SSR, STS	Bimpong et al. (2011)
Productive panicle number	*O. rufipogon*	G52-9	Yuexiangzhan	AB-QTL	CIM	2, 3, 7	SSR	Jing et al. (2010)
Spikelets per plant	*O. minuta*	IRGC101144	Hwaseongbyeo	NIL		7	SSR	Balkunde et al. (2013)
	O. rufipogon	G52-9	Yuexiangzhan	AB-QTL	CIM	2	SSR	Jing et al. (2010)
	O. rufipogon	IC 22015	IR 58025A	AB-QTL	IM, CIM	2, 5	SSR	Marri et al. (2005)
	O. rufipogon	IRGC105491	V20A, V20B	B_C2	ANOVA	1	RFLP	Xiao et al. (1998)
	O. rufipogon	IRGC105491	MR219	AB-QTL	CIM	1	SSR	Wickneswari et al. (2012)
Spikelets per panicle	*O. glaberrima*	IRGC103544	Milyang 23	$BC_{3F}2$	SPA	3	SSR	Suh et al. (2005)
	O. grandiglumis	IRGC101144	Hwaseongbyeo	AB-QTL	SPA	2, 3, 4, 11	SSR	Yoon et al. (2006)
	O. minuta	IR71033-121-15	Junambyeo	$_{F2:}3$	SPA, IM	6	SSR, STS	Rahman et al. (2007)
	O. rufipogon	IRGC105491	Hwaseongbyeo	NIL, AB-QTL	SPA, IM, ANOVA	8, 9	SSR	Xie et al. (2008), Jin et al. (2009)
	O. rufipogon	IRGC105491	Jefferson	AB-QTL	SPA, IM, CIM	2, 3, 9	RFLP, SSR	Thomson et al. (2003)
	O. rufipogon	IRGC105491	V20A, V20B	B_C2	ANOVA	1, 9	RFLP	Xiao et al. (1998)
	O. rufipogon	W1944	Hwayeongbyeo	IL, RIL	SPA, IM, CIM	1	SSR	Yuan et al. (2009), Lee et al. (2005)
	O. rufipogon	YJCW	93-11	AB-QTL	SPA, IM, CIM	3	SSR	Fu et al. (2010)
	O. glaberrima		Milyang 23	$BC_{2F}5$		2	SSR	Kang et al. (2008)

	O. rufipogon	IRGC105491	MR219	AB-QTL	CIM	3	SSR	Wickneswari et al. (2012)
	O. minuta	IRGC101144	Hwaseongbyeo	NIL		7	SSR	Balkunde et al. (2013)
Spikelet fertility	*O. glaberrima*	IRGC103544	Milyang 23	$BC_{3F}2$	SPA	2, 4, 8	SSR	Suh et al. (2005)
	O. longistaminata		RD23	$BC_{7F}2$	CIM	6	SSR	Chen et al. (2009)
	O. minuta	IR71033-121-15	Junambyeo	$_{F2:}3$	SPA, IM	6	SSR, STS	Rahman et al. (2007)
Shattering	*O. rufipogon*	IRGC105491	Jefferson	AB-QTL	SPA, IM, CIM	8	RFLP, SSR	Thomson et al. (2003)
	O. rufipogon	W1944	Hwayeongbyeo	RIL	SPA, CIM	1, 3, 6	SSR	Lee et al. (2005)
	O. rufipogon	W1944	Hwayeongbyeo	IL, RIL	SPA, IM, CIM	1, 4, 5	SSR	Yuan et al. (2009), Lee et al. (2005)
Grains per panicle	*O. minuta*	IRGC101144	Hwaseongbyeo	NIL		7	SSR	Balkunde et al. (2013)
	O. rufipogon	IRGC105491	Jefferson	AB-QTL	SPA, IM, CIM	2, 3, 8, 9	RFLP, SSR	Thomson et al. (2003)
	O. rufipogon	IRGC105491	V20A, V20B	B_C2	ANOVA	1, 8, 12	RFLP	Xiao et al. (1998)
	O. rufipogon	IRGC105491	Hwaseongbyeo	AB-QTL, NIL	SPA, IM, ANOVA	8, 9	SSR	Jin et al. (2009), Xie et al. (2008)
	O. rufipogon	IC22015	IR 58025A	AB-QTL	IM, CIM	2, 5	SSR	Marri et al. (2005)
	O. rufipogon	G52-9	Yuexiangzhan	AB-QTL	CIM	4, 10, 11	SSR	Jing et al. (2010)
	O. rufipogon	YJCW	93-11	AB-QTL	SPA, IM, CIM	1, 3	SSR	Fu et al. (2010)
Percent seed set	*O. meyeriana*	Y73	IR24	RIL	CIM	8	SSR, STS	Chen et al. (2012)
	O. rufipogon	IRGC105491	MR219	AB-QTL	CIM	3	SSR	Wickneswari et al. (2012)
	O. rufipogon	IRGC105491	V20A, V20B	B_C2	ANOVA	2, 4	RFLP	Xiao et al. (1998)

	O. rufipogon	IRGC105491	Hwaseongbyeo	NIL	SPA, IM, ANOVA	8	SSR	Xie et al. (2006)
	O. rufipogon	IRGC105491	Jefferson	AB-QTL	SPA, IM, CIM	1, 5	RFLP, SSR	Thomson et al. (2003)
	O. rufipogon	IRGC105491	IR64	AB-QTL	SPA, IM, CIM	1, 3	RFLP, SSR	Septiningsih et al. (2003)
	O. rufipogon	IRGC105491	V20A, V20B	B_C2	ANOVA	4, 8, 9, 11, 12	RFLP	Xiao et al. (1998)
	O. rufipogon	IRGC105491	MR219	AB-QTL	CIM	6	SSR	Wickneswari et al. (2012)
	O. rufipogon	IRGC105491	Hwaseongbyeo	NIL	ANOVA	9	SSR	Xie et al. (2008)
	O. rufipogon	IRGC105491	Hwaseongbyeo	AB-QTL	SPA, IM	8	SSR	Jin et al. (2009)
	O. rufipogon	IRGC105491	Ce64&V20A, Caiapó, Hwacheong, Jefferson, IR64	AB-QTL	IM, CIM	3	SSR	McCouch et al. (2007)
	O. rufipogon	IC22015	IR 58025A	AB-QTL	IM, CIM	2, 9	SSR	Marri et al. (2005)
	O. rufipogon	W1944	Hwayeongbyeo	IL	SPA, IM	1	SPA, IM	Yuan et al.(2009)
	O. rufipogon	YJCW	93-11	AB-QTL	SPA, IM, CIM	1	SSR	Fu et al. (2010)
Brown rice yield	*O. glaberrima*	IRGC103544	V20A	AB-QTL	SPA, IM, CIM	12	RFLP, SSR	Li et al. (2004)
Grain yield per plant	*O. glaberrima*		IR64	$BC_{2F}3$	SPA	2, 6, 8, 9	SSR, STS	Bimpong et al. (2011)
	O. glaberrima	IRGC103544	V20A	AB-QTL	SPA, IM, CIM	12	RFLP, SSR	Li et al. (2004)
	O. rufipogon		MR219	AB-QTL		1	SSR	Bhuiyan et al. (2011)
	O. rufipogon	G52-9	Yuexiangzhan	AB-QTL	CIM	1, 2, 3		Jing et al. (2010)
	O. rufipogon	IC22015	IR 58025A	AB-QTL	IM, CIM	2, 9	SSR	Marri et al. (2005)

	O. rufipogon	IRGC105491	IR64	AB-QTL	SPA, IM, CIM	1	RFLP, SSR	Septiningsih et al. (2003)
	O. rufipogon	IRGC105491	Hwaseongbyeo	NIL	ANOVA	9	SSR	Xie et al. (2008)
	O. rufipogon	IRGC 105491	V20A, V20B	B_C2	ANOVA	1, 2, 8	RFLP	Xiao et al. (1998)
Yield	*O. glaberrima*	IRGC103544	Milyang 23	$BC_{3F}2$	SPA	2, 3, 4, 6, 8	SSR	Suh et al. (2005)
	O. glaberrima	IRGC103544	Milyang 23	$BC_{2F}5$		2	SSR	Kang et al. (2008)
	O. grandiglumis	IRGC101154	Hwaseongbyeo	AB-QTL	SPA	2	SSR	Yoon et al. (2006)
	O. minuta	IRGC101144	Hwaseongbyeo	NIL		7	SSR	Balkunde et al. (2013)
	O. meyeriana	Y73	IR24	RIL	CIM	6	SSR, STS	Chen et al. (2012)
	O. rufipogon		MR219	AB-QTL		4	SSR	Bhuiyan et al. (2011)
	O. rufipogon	IC22015	IR58025A	AB-QTL	IM, CIM	1, 2, 8	SSR	Marri et al. (2005)
	O. rufipogon	IRGC105491	Jefferson	AB-QTL	SPA, IM, CIM	2, 3, 6, 9	RFLP, SSR	Thomson et al. (2003)
	O. rufipogon	IRGC105491	IR64	AB-QTL	SPA, IM, CIM	1	RFLP, SSR	Septiningsih et al. (2003)
	O. rufipogon	IRGC105491	V20A, V20B	B_C2	ANOVA	1, 2, 8, 12	RFLP	Xiao et al. (1998)
	O. rufipogon	IRGC105491	Hwaseongbyeo	AB-QTL	SPA, IM	8	SSR	Jin et al. (2009)
	O. rufipogon	YJCW	93-11	AB-QTL	SPA, IM, CIM	1	SSR	Fu et al. (2010)
Harvest index	*O. glaberrima*		IR64	$BC_{2F}3$	SPA	2, 7	SSR, STS	Bimpong et al. (2011)
	O. rufipogon	IC22015	IR58025A	AB-QTL	IM, CIM	2	SSR	Marri et al. (2005)

[i] - † Abbreviations for mapping population types are: AB-QTL, advanced backcross-quantitative trait locus; IL, inbred line; NIL, near isogenic line; RIL, recombinant inbred line.

[ii] - ‡ Abbreviations for QTL analysis method are: ANOVA, analysis of variance; CIM, composite interval mapping; IM, interval mapping; MIM, multiple interval mapping; SPA, single point analysis.

[iii] - § Only the chromosomes where the QTL increase is attributed to the wild parent are listed.

[iv] - Abbreviations for types of markers are: RAPD, random amplified polymorphic DNA; RFLP, restriction fragment length polymorphism; SNP, single nucleotide polymorphism; SSR, simple sequence repeat; STS, sequence-tagged site.

Yield Enhancing QTL from Exotic Oryza Genomes

Several plant traits directly or indirectly affect rice grain yield including days to heading and maturity; plant height; panicle length; number of panicles per plant, spikelets per panicle and grains per panicle; seed set; grain weight; grain size and shape; and shattering (Table 2). The most important yield components in rice are the number of panicles per plant, number of grains per panicle, and grain weight (Chen et al. 2012; Lee et al. 2004; Septiningsih et al. 2003; Thomson et al. 2003). Yield improvement can be achieved as a result of the vast allelic diversity for these traits found in interspecific populations, especially number of grains per panicle which has proven to have the greatest relevance for rice breeding programs (Li et al. 1998; Liu et al. 2008; Tian et al. 2006).

Modern rice varieties are developed after an extensive selection process to improve a few targeted traits related to cultivation and end-use quality but primarily those associated with yield components, such as resistance to shattering, compact growth habit and improved seed germination (Tanksley and McCouch 1997). This prolonged breeding procedure can lead to a reduction in the genetic variability found in modern cultivated rice (Rangel et al. 2008), thus identifying genetic sources for agronomically important traits from wild *Oryza* species and introgressing them into cultivated rice is desirable and necessary.

Although wild *Oryza* species are inferior in grain yield, especially when compared to cultivated rice, transgressive segregation resulting from a cross between cultivated rice and a wild*Oryza* species, especially the ancestral species, *O. rufipogon* and *O. nivara,* revealed the presence of favorable alleles from the wild parent that can increase yield in the genetic background of cultivated rice (Brar and Singh 2011). Xiao et al. (1996) developed a genetic population by initially crossing the cytoplasmic male sterile parent, V20A, with *O. rufipogon* (IRGC105491), the donor, as male parent. Subsequently, F_1 plants and later BC_1 plants selected for vigor, plant type, maturity and fertility, were backcrossed to V20B (maintainer line of V20A). A selected subset of 300 BC_2F_1 lines was crossed with Ce64 to create the genotype of the Chinese hybrid rice variety V/64, developed from V20A x Ce64. Xiao et al. (1996) reported that 15% of the testcross families outperformed the recurrent parent, 14% of the yield improvement was related to grains per plant and 56% was related to 1000-grain weight. Subsequently, Xiao et al. (1998) identified 35 QTLs associated with yield improvement, 19 of which, including *yld1.1* and *yld 2.1,* were located on chromosomes 1 and 2, respectively. They also observed no undesirable alleles causing negative effects on yield components, thus the presence of alleles in wild *O. rufipogon* can improve rice yields.

Other AB-QTL populations developed using the same *O. rufipogon* (IRGC105491) donor parent with recurrent parents representing various *O. sativa* subpopulations including *indica* with IR64 (Septiningsih et al. (2003), upland *tropical japonica* with Caiapó (Moncada et al. 2001), irrigated*tropical japonica* with Jefferson (Thomas et al. 2003), and *temperate japonica* with Hwaseongbyeo (Xie et al. 2006 & 2008), revealed enhanced yield and yield components attributed to the *O. rufipogon*donor parent. Selected progeny lines were advanced to BILs or NILs and this yield enhancement was confirmed in field studies with the recurrent parents IR64 (Cheema et al. 2008a), Jefferson (Imai et al. 2013), and Hwaseong byeo (Jin et al. 2011). Also, an epistatic interaction was noted between the QTLs for grain weight on chromosomes 8 and 9 in the Hwaseongbyeo background (Jin et al. 2011).

Tian et al. (2006) selected an introgression line, SIL040, from the BC_4F_4 lines of *O. rufipogon*(Dongxiang) x Guichao 2, an *indica* rice cultivar. High resolution QTL mapping and analysis in the SIL040 x Guichao 2 F_2 progeny for yield components revealed *gpa7* (grains

5 (Hu et al. 2011; Nagao and Takahashi 1963; Takamure and Kinoshita 1991). *O. meridionalis* has long awns, ranging in length from 7.8-10.3 cm (Vaughan 1994) and two genes, *An7*and *An8*, associated with the trait, were identified on chromosomes 5 and 4, respectively (Kurakazu et al. 2001). Analysis of *O. meridionalis* x *O. sativa* $BC_4F_{2:6}$ and $BC_4F_{2:8}$ revealed the presence of two dominant genes at different locations on chromosome 1, *An9* and *An10*, and a new allele, *An6-mer* on chromosome 6 (Matsushita et al. 2003a). Another study of an *O. sativa* x *O. glumaepatula* population also identified new alleles, *An7* and *An8* (Matsushita et al. 2003b), thus confirming awn length is controlled by several genes.

A doubled haploid (DH) population was developed from Caiapó (*tropical japonica*, recurrent parent) x*O. glaberrima* (donor parent, IRGC103544, MG12) BC_3F_1 progeny (Aluko et al. 2004). This population was evaluated for agronomic traits including yield and yield components in Colombia and Louisiana, USA (Gutiérrez et al. 2010). Strong segregation distortion was found on chromosomes 3 and 6 indicating the presence of interspecific sterility genes. Evaluation of the phenotypic data revealed transgressive segregation for several traits. A set of 34 CSSLs was selected from Koshihikari, an elite*temperate japonica* rice cultivar (recurrent parent) x *O. glaberrima* (donor parent, IRGC104038) BC_5F_1 progeny, advanced to the F_7 generation, and genotyped with 142 SNP markers (Shim et al. 2010). QTL analysis of phenotypic data from field grown plants revealed 105 putative QTL of which 84 were positive with 64 being related to grain yield components, suggesting the possible use of favorable alleles in *O. glaberrima* for improvement of cultivated rice.

These studies give several examples of QTL or genes for yield and yield components being attributed to the wild donor parent not only the ancestral A-genome species, *O. rufipogon* or *O. nivara*, but also in the more distant tetraploid *O. minuta* with a BBCC genome. These observations confirm that not only single genes and alleles are affecting these traits but there are epistatic interactions and epigenetic interactions, as well as environmental factors affecting many of these traits, resulting in the phenomenon often described as transgressive variation. Currently, six CSSL libraries are under development with three different *O. rufipogon/O. nivara* donor accessions, representing the *Indica*,*Japonica* and independent groups of *O. rufipogon* (Huang et al. 2012), and two recurrent parents, IR64, an *indica* representing the *Indica* subspecies, and Cybonnet, a U.S. *tropical japonica*, representing the *Japonica* subspecies to further explore these interactions resulting in transgressive variation (Tung et al. 2010; SR McCouch, Cornell University and GC Eizenga, personal communication).

GENES FOR STERILITY

Reproductive barriers, such as crossability, hybrid seed inviability, hybrid sterility and hybrid breakdown, have significantly limited the success of interspecific hybridization between *O. sativa* and non-A genome *Oryza* species. Several studies reported the production of F_1 seeds by crossing male sterile lines and *Oryza* species (Huang et al. 2001; Luo et al. 2000). The crossability rate between *O. sativa* and other *Oryza* species vary; however, the rate of crossability between A-genome and non-A genome diploid *Oryza* species is higher than with tetraploid *Oryza* species, none of which has an A-genome (Jena and Khush 1989 & 1990; Yasui and Iwata 1991).

The phenomenon of transmission ratio distortion (TRD) where one allele is transmitted more frequently than the opposite allele in interspecific and intraspecific hybrids has been discovered in a broad range of organisms and is often a reproductive barrier (Koide et al. 2012). Recently, Koide et al. (2012) identified a unique sex-independent TRD (*si*TRD) where one allele is preferentially transmitted through both the male and female parent derived from *O. rufipogon* (W593). This research showed the S_6 allele on chromosome 6 is responsible for the *si*TRD allele and influenced by other unlinked modifiers. The locus, *sa1*, conferring pollen sterility was found in *O. glaberrima* (W025) x T65*wx* progeny (Sano 1990) where T65*wx* is an NIL derived from Taichung 65 (*japonica* rice) x Peiku (*indica* rice) with a Taichung 65 background and the Peiku *waxy* gene on chromosome 6. Other studies identified several pollen sterility loci, *S-1, S3, S18, S19, S20, S21, s25, s27, S29, S29(t)* and *s36*, in populations resulting from interspecific hybridizations between various *O. sativa* cultivars and the *Oryza* species, *O. glumaepatula*, *O. glaberrima* and *O. nivara* (Doi et al. 1998 & 1999; Hu et al. 2006, Sano 1983 & 1986; Taguchi et al. 1999; Win et al. 2009). To overcome such barriers, several methods have been suggested including anther culture, backcrossing, marker-assisted selection (MAS) and asymmetric somatic hybridization, (Fu et al. 2008; Sarla et al. 2005). Also, Deng et al. (2010) demonstrated the fertility in *O. glaberrima* x *O. sativa* crosses could be improved by using an *O. sativa* bridging parent. The bridging parent had the *O. glaberrima* sterility gene, *S1-g* on chromosome 6, introgressed into the particular *O. sativa* cultivar background.

GRAIN QUALITY TRAITS

Acceptable rice grain quality is a major goal of rice breeding programs worldwide because it determines the acceptability of cooked rice to the consumer. Grain quality is a combination of several components including milling efficiency, physical appearance, cooking and eating characteristics, and nutritional quality (Aluko et al. 2004; Li et al. 2004). A few interspecific populations were evaluated for grain quality traits (Table 3). These studies showed the *Oryza* parent affects the apparent amylose content, alkali spreading value, protein content, rice bran percentage, milled rice percentage and seed size. What is desirable for these traits is determined for the most part, by consumer preference and marketing classes. When selecting for these traits, often the grain quality of the recurrent parent is preferred.

Most interesting was the BC_3F_1 progeny of the Caiapó x *O. glaberrima* (IRGC103544, MG12) doubled haploid population (Aluko et al. 2004). For this population, the QTL analysis revealed 27 QTLs associated with rice quality of which seven QTLs including percent rice bran, percent milled rice, alkali spreading value (inversely related to gelatinization temperature), percent protein and grain dimensions (length to width ratio), were traced to alleles originating from the *O. glaberrima* parent.

Trait	Donor species	Donor accession	Recurrent parent(s)	Mapping population†	QTL mapping analysis‡	Chromosome location§	Type of marker¶	References
Grain Quality								
Apparent amylose content	*O. glaberrima*	IRGC103544	V20A	AB-QTL	SPA, IM, CIM	6, 12	RFLP, SSR	Li et al. (2004)
O. grandiglumis	IRGC101154	Hwaseong-byeo	AB-QTL	SPA	3, 5, 7	SSR	Yoon et al. (2006)	
	O. nivara	IRGC81848	Swarna	$BC_{2F}2$	IM, CIM	2	SSR	Swamy et al. (2012)
	O. nivara	IRGC100195	M-202	AB-QTL	MIM	6	SSR	Eizenga et al (accepted)
	O. rufipogon	IRGC105491	IR64	AB-QTL	IM, CIM	6	RFLP, SSR	Septiningsih et al. (2003b)
Alkali spreading value (ASV) or gel consistency	*O. glaberrima*	IRGC103544	Caiapó	$BC_{3F}1$	IM, CIM	6	SSR	Aluko et al. (2004)
O. glaberrima	IRGC103544	V20A	AB-QTL	SPA, IM, CIM	12	RFLP, SSR	Li et al. (2004)	
	O. nivara	IRGC100195	M-202	AB-QTL	MIM	6	SSR	Eizenga et al. (accepted)
	O. rufipogon	IRGC105491	IR64	AB-QTL	IM, CIM	6	RFLP, SSR	Septiningsih et al. (2003b)

Kernel elongation	*O. glaberrima*	IRGC 103544	V20A	AB-QTL	SPA, IM, CIM	3	RFLP, SSR	Li et al. (2004)
Protein	*O. glaberrima*	IRGC103544	Caiapó	$BC_{3F}1$	IM, CIM	2, 6	SSR	Aluko et al. (2004)
	O. glaberrima	IRGC103544	V20A	AB-QTL	SPA, IM, CIM	8	RFLP, SSR	Li et al. (2004)
Percent rice bran	*O. glaberrima*	IRGC103544	Caiapó	$BC_{3F}1$	IM, CIM	4, 7	SSR	Aluko et al. (2004)
Percent milled rice	*O. glaberrima*	IRGC103544	Caiapó	$BC_{3F}1$	IM, CIM	5	SSR	Aluko et al. (2004)
	O. nivara	IRGC81848	Swarna	$BC_{2F}2$	IM, CIM	1	SSR	Swamy et al. (2012)
Biotic Stress-Diseases								
Bacterial blight	*O. australiensis*		IR31917-45-3-2	MAAL		12		Multani et al. (1994)
	O. brachyantha O. longistaminata O. officinalis			AIL		5, 6, 8, 11	SSR	Hechanova et al. (2008)
	O. latifolia	IRGC100914	IR31917-45-3-2	AIL	ANOVA	12, others	SSR, SNP, STS, InDel	Angeles-Shim et al. (accepted)
	O. longistaminata	WLO2	BS125	NIL		11	RFLP, RAPD	Ronald et al. (1992)
	O. longistaminata		IR24			11		Khush et al. (1990)

	O. meyeriana	Y73	IR24	RIL	CIM	1, 3, 5, 10, 11	SSR, STS	Chen et al. (2012)
	O. minuta	78-1-5	IR24	F_2- B_C1		6	RAPD, AFLP	Gu et al. (2004)
	O. nivara	IRCC81825	PR114	RIL, BIL, IL	SMA-IM	4	SSR, STS	Cheema et al. (2008)
Blast disease	*O. australiensis*	IRCC100882	Lijiangxintuan-heigu			6	CAPS, SSR, STS	Jeung et al. (2007)
	O. minuta	IRGC101141	IR31917	$_F2$		6	RAPD	Liu et al. (2002)
	O. nivara	IRGC100898	Bengal	AB-QTL	MIM	8	SSR	Eizenga et al. (2013)
	O. nivara	IRGC104705	Bengal	AB-QTL	MIM	8, 12	SSR	Eizenga et al. (2013)
	O. rufipogon	IRGC104812	Koshihikari	IL		3, 11		Hirabayashi et al. (2010); Sobrizal et al. (1999)
	O. rufipogon	IRGC104814	Koshihikari	IL		3, 5, 6		Hirabayashi et al. (2010)
Sheath blight disease	*O. nivara*	IRGC100898	Bengal	AB-QTL	MIM	6	SSR	Eizenga et al. (2013)
	O. nivara	IRGC104705	Bengal	AB-QTL	MIM	3, 6	SSR	Eizenga et al. (2013)
Stem rot	*O. rufipogon*	IRGC100912 (87-Y-550)	L-201 (long grain-breeding lines)	$_F2$	SPA	2, 3	AFLP	Ni et al. (2001)

Grassy stunt virus	*O. nivara*	IRGC101508	IR8, IR20, IR22, IR24, IR773A-1-3	$_{F}2$				Nuque et al. (1982)
Rice stripe necrosis resistance	*O. glaberrima*	IRGC103544	Caiapó	CSSL	IM, CIM	11	SSR	Gutiérrez et al. (2010)
Biotic Stress-Insects								
Brown planthopper	*O. australiensis*	IRGC100882	IR31917-45-3-2	IL		12	RFLP	Ishii et al. (1994)
	O. australiensis	IR65482-7-216-1-2	Jinbubyeo	F_2, $BC_{2F}2$	ANOVA	12	SSR, STS	Jena et al. (2006)
	O. australiensis		IR31917-45-3-2	MAAL		12		Multani et al. (1994)
	O. eichingeri	IRGC105159	2428	F_2, $BC_{1F}1$		2	RFLP, SSR	Guoqing et al. (2001)
	O. latifolia	B14	Taichung	RIL		4	SSR, RFLP	Yang et al. (2002)
	O. minuta	101141	Junambyeo	$_{F}3$		4, 12	SSR, STS	Rahman et al. (2009)
	O. officinalis	IRGC100878, IRGC100896, IRGC101150, IRGC101412, IRGC102385	IR31917-45-3, IR25587-109-3	$BC_{2F}8$		4, 10, 12	RFLP	Jena et al. (1992)
	O. officinalis	IR54745-2-21-12-17-6	IR50			3	RAPD, STS	Renganayaki et al. (2002)
	O. officinalis	B5	1826, 93-11			3, 4	SSR	Li et al. (2006)

	O. officinalis	IRGC100896	IR31917-45-3-2	F_2		11	RAPD	Jena et al. (2002)
	O. officinalis					3	RFLP	Hirabayshi et al. (1998)
	O. officinalis	IR54745-2-21-12-17-6	IR50	RIL		3	RAPD	Renganayaki et al. (2002)
	O. officinalis	B5	Minghui 63	F_2		3	RFLP	Huang et al. (2001)
	O. officinalis	B5		RIL		4	AFLP, RFLP, SSR	Yang et al. (2004)
	O. rufipogon	WBO1	Minghui 63	F_2		4, 8	SSR	Hou et al. (2011)
Green rice leafhopper	*O. glaberrima*	IRGC104038	Taichung 65	NIL	IM, CIM	3, 7, 9, 10	SSR	Fujita et al. (2010)
	O. rufipogon	W1962	Taichung 65	NIL, BC_4F_2		8	SSR	Fujita et al. (2006)
White-backed planthopper	*O. officinalis*	B5	Minghui 63	RIL		3, 4	SSR	Tan et al. (2006b)
	O. rufipogon	BILs-DWR/Dongxiang	XieqingzaoB	BIL	CIM, MIM	2, 5, 9	SSR	Chen et al. (2010)
Abiotic stress								
Aluminum tolerance	*O. rufipogon*	IRGC106424	IR64	RIL	IM	1, 3, 9 (2, 7, 8)	RFLP	Nguyen et al. (2003)
Drought tolerance	*O. rufipogon*		Guichao 2	IL	SMR	2, 12	SSR	Zhang et al. (2006)

	O. rufipogon	W630	Nipponbare	BIL	IM	1, 5	SSR	Thanh et al. (2011)
Heat tolerance	*O. rufipogon*	YJCWR	TeQing	IL	CIM	1	SSR	Lei et al. (2013)
Low temperature tolerance	*O. nivara*	IRGC100195	M-202	AB-QTL	MIM	5	SSR	Eizenga et al. (accepted)
	O. rufipogon	W1943	Guang-lu-ai 4	$BC_{4F}2$	IM, CIM	3, 11	SNP	Koseki et al. (2010)
	O. rufipogon	Dongxiang	Nanjing11	$BC_{2F}1$	CIM	10	SSR	Xia et al. (2010)
	O. rufipogon	W1944	Hwayeong-byeo	RIL	SPA, CIM	2, 5	SSR	Lee et al. (2005)
Salt tolerance	*O. rufipogon*		TeQing	IL	SPA	1, 2, 3, 6, 7, 10	SSR	Tian et al. (2011)
Submergence stress	*O. rufipogon*					9		Li et al. (2011)
Biomass	*O. glaberrima*		IR64	$BC_{2F}3$	SPA	1, 2, 3, 6, 10	SSR, STS	Bimpong et al. (2011)

[i] - † Abbreviations for mapping population types are: AB-QTL, advanced backcross-quantitative trait locus; AIL, alien introgression line; BIL, backcrossed inbred line; CSSL, chromosome segment substitution line; IL, inbred line; MAAL, monosomic alien addition line; NIL, near isogenic line; RIL, recombinant inbred line.

‡ Abbreviations for QTL analysis method are: ANOVA, analysis of variance; CIM, composite interval mapping; IM, interval mapping; MIM, multiple interval mapping; SMR, single marker analysis; SPA, single point analysis.

[ii] - § Only the chromosomes where the QTL increase is attributed to the wild parent are listed.[iii] -

Abbreviations for types of markers are: AFLP, amplified fragment length polymorphism; CAPS, cleaved amplified polymorphic sequence, InDel, insertion-deletion polymorphism, RAPD, random amplified polymorphic DNA; RFLP, restriction fragment length polymorphism; SNP, single nucleotide polymorphism; SSR, simple sequence repeat; STS, sequence-tagged site.

RESISTANCE TO BIOTIC STRESS

Disease Resistance

Pathogenic microorganisms, such as fungi, oomycetes, viruses and bacteria, and pests, such as insect herbivores, significantly reduce rice seed yield and quality. The most destructive rice diseases include bacterial blight caused by *Xanthomonas oryzae* pv. *oryzae* Ishiyama Dye (Cheema et al. 2008), blast caused by the fungus *Magnaporthe oryzae* B. Couch (Couch and Kohn 2002), and sheath blight caused by the soil-borne fungus *Rhizoctonia solani* Kühn (Zhang 2007). The first reported successful introduction of an agronomically important trait from a wild *Oryza* species was the introgression of grassy stunt virus resistance from the AA-genome species *O. nivara* into the cultivated *O. sativa*genetic background (Khush et al. 1977). The resistance mechanism was subsequently introduced into several other rice cultivars (Sanchez et al. 2013). Since this first introduction, wild *Oryza* accessions have been screened as a potential source of resistance genes to biotic and abiotic stresses, as well as, yield and yield components, as previously discussed. These screening efforts, including successful introduction of stress resistance genes from *Oryza* species were recently summarized by Ali et al. (2010), Brar and Singh (2011) and Sanchez et al. (2013). Table 3 summarizes the efforts to identify the chromosome location of stress resistance genes introduced from the wild *Oryza* species by QTL and fine mapping analyses.

Seed yield losses due to bacterial blight were reported to be up to 75% in India, Indonesia, and the Philippines, and 20 to 30% in Japan (Mew et al. 1993; Nino-Liu et al. 2006). Thus far 31 bacterial blight resistance genes have been reported and six of these were identified in species other than *O. sativa*. These resistance genes were identified in several wild *Oryza* species, including *Xa21* in *O. longistaminata*, *Xa23* in *O. rufipogon*, *Xa27* in *O. minuta* (IRGC101141), *Xa29*(t) in *O. officinalis*(B5), and *Xa30*(t) in *O. nivara* (IRGC81825) (Brar and Singh 2011; Cheema et al. 2008b; Gu et al. 2004; Khush et al. 1990; Tan et al. 2004; Zhang et al. 1998). Most recently, *Xa34*(t) exhibiting broad spectrum resistance, was identified in *O. brachyantha* (IRGC101232) as a single dominant gene after examining the crossed progeny of two resistant introgression lines derived from IR56 (recurrent parent) and *O. brachyantha* (Ram et al. 2010a).

Both bacterial blight and blast resistance were identified in the tetraploid CCDD genome species, *O. minuta* (IRGC101141). To transfer these resistance genes into the background of diploid, cultivated rice, Amante-Bordeos (1992) used embryo rescue and backcrossing to produce interspecific hybrids between the elite *O. sativa* line, IR31917-45-3-2 (recurrent parent) and *O. minuta* (donor parent).

Lastly, the line Y73 was selected for a high level of bacterial blight resistance from the hybrid progeny of an asymmetric somatic hybridization between a resistant *O. meyeriana* and a *O. sativa* subsp. *japonica* cultivar (Yan et al. 2004). Subsequently, Chen et al. (2012) developed a RIL population from Y73 x IR24 (*O. sativa*) and identified five QTLs, two were major QTLs located on chromosomes 1 and 5, and three were minor QTLs on chromosomes 3, 10 and 11. They also mapped 21 additional QTLs related to yield components and yield.

Blast is considered the most destructive fungal disease in irrigated rice. The symptoms include lesions on leaves, stems, peduncles, panicles, seeds and roots (Savary and Willocquet 2000; Khush et al. 2009). To date, 41 blast resistance genes have been reported; however, there are only two genes, *Pi9*and *Pi40(t)*, that have been identified in wild *Oryza* species, *O. minuta* and *O. australiensis*, respectively. *Pi40(t)*, which confers broad spectrum of blast resistance, was introgressed into the elite breeding line, IR65482-4-136-2-2 (Liu et al. 2002; Jeung et al. 2007). Qu et al. (2006) cloned the *Pi9*gene via a positional (map-based) cloning technique and found the gene is a member of a group of six resistance-like genes, which encodes a nucleotide-binding site (NBS) and leucine-rich repeats (LRRs).Silué et al. (1992) screened 99 *O. glaberrima* accessions for blast resistance by inoculating with ten *M. oryzae* strains from Cote d›Ivoire. The results revealed that nine accessions were resistant to all *M. oryzae* strains and

32 accessions were moderately resistant, suggesting these accessions may be the source of novel resistance genes. Eizenga et al. (2009) identified resistance to U.S. blast races in some*Oryza* species but no accession exhibited resistance to all races. Subsequently, two AB-QTL populations with two different resistant *O. nivara* (IRGC100898; IRGC104705) accessions as donor parents x Bengal, a U.S. medium grain *tropical japonica*, as recurrent parent were evaluated for reaction to two U.S. blast races. QTL analysis of these populations mapped resistance to U.S. leaf blast races *IB1* and *IB49* to chromosomes 8 and 12 (Eizenga et al. 2013).

Rice sheath blight, *Rhizoctonia solani* Kühn, was reported for the first time in Japan in 1910 and since then, it has spread to many rice growing areas worldwide (Lee and Rush 1983; Savary et al. 2000). To date, no major resistance gene(s) that confers complete resistance to sheath blight has been discovered, only genes conferring partial resistance (Pinson et al. 2005; Jia et al. 2009). Several *Oryza*species accessions were screened for sheath blight resistance at the International Rice Research Institute (IRRI) in the Philippines with resistance being identified in accessions of *O. minuta* and *O. rufipogon* (Amante et al. 1990), and a resistant *O. officinalis* accession being the donor of sheath blight resistance genes in *O. sativa* introgression lines (Lakshmanan, 1991). Prasad and Eizenga (2008) screened a collection of 73 *Oryza* species accessions using three different methods and identified seven accessions with moderate resistance including three *O. nivara* accessions and one each of *O. barthii*,*O. meridionalis*, *O. nivara*/*O. sativa* hybrid, and *O. officinalis*. Based on these results, Eizenga et al. (2013) developed two AB-QTL populations using two of these *O. nivara* accessions (IRGC100898; IRGC104705) as the donor parents, and Bengal as the recurrent parent for both populations. QTL analysis identified a major QTL, *qShB6*, associated with sheath blight attributed to the wild rice parent. Two other minor QTLs, *qShB1* and *qShB3*, were identified but not attributed to the *O. nivara* parent in all sheath blight tests. A third AB-QTL population with the more distant A-genome species, *O. meridionalis* (IRGC105306), as the donor parent, and Lemont, a U.S. long grain *tropical japonica*, as the recurrent parent, was evaluated for reaction to sheath blight disease and the QTL analysis is currently underway (Eizenga, unpublished data).

Stem rot, a fungal disease caused by *Sclerotium oryzae* Catt., causes yield losses by reduced tillering, more unfilled grains per

from the soil colloids and come into a solution phase. Aluminum at toxic levels slows root development, reduces the plant's ability to take up water and nutrients, and decreases plant growth, consequently reducing grain yield and grain quality (Foy 1992). Application of lime to the soil, reduces soil acidity and improves soil fertility but the results have showed limited success in overcoming the effects of aluminum toxicity. Aluminum tolerance is a quantitative trait and varies among rice species. Both additive and dominance effects contribute to the genetic heritability of aluminum tolerance as documented by the importance of both general combining ability and specific combining ability (Howeler and Cadavid 1976; Wu et al. 1997).

In the past decade, one *O. rufipogon* (IRGC106424) accession found growing in an acid sulfate soil in Vietnam (Sanchez et al. 2013) has proven to be valuable for improving tolerance to both aluminum and acid sulfate soils in cultivated rice. Initially, Nguyen et al. (2003) evaluated 171 F_6 RILs derived from IR64 (*indica*, susceptible) x *O. rufipogon* (IRGC106424, tolerant) for aluminum tolerance. QTL analysis revealed QTLs for root length under stress conditions attributed to the *O. rufipogon* parent in six different chromosomal regions on chromosomes 1, 2, 3, 7, 8 and 9 that individually explained 9.0–24.9% of the phenotypic variation and were controlled by additive effects. The major QTL on chromosome 3, explaining 24.9% of the variation, was found to be conserved across cereal species. During the same time period, the tolerance to acid sulfate soils identified in this *O. rufipogon* accession was introgressed into the IR64 background through breeding efforts. The selected introgression line, IR73678-6-9-B, was released by IRRI as variety AS996 (Sanchez et al. 2013). AS996 is currently grown on 100,000 ha in the Mekong Delta and described as moderately tolerant to acid sulfate soils and tolerant to brown planthopper and blast. Even though traits associated with abiotic stress are more difficult to evaluate because of environmental effects and interactions between genes, the development of the AS996 variety is an exciting success story. The release of Arizona-1 and Arizona-2 could make significant contributions to improving rice yields in areas where high temperatures routinely lower yield. With the improved molecular techniques for dissecting these traits and the gene functions related to abiotic stress, more significant advances should be made in the near future, especially as the scientific community provides the tools for rice producers to deal with global climate change.

CONCLUSIONS

The repositories of *Oryza* species accessions found around the world are a storehouse of novel alleles and traits lost during the evolution and domestication of cultivated rice as we know it today. The fact that introgression lines derived from crosses between Asian rice and it's ancestral species, *O. rufipogon* and *O. nivara*, exhibited notable improvement in yield and yield components through the phenomenon known as transgressive variation, was surprising and unexpected. The identification of novel alleles related to biotic stress, especially insect pests like brown planthopper and bacterial leaf blight, and more recently abiotic stresses like acid sulfate soils and drought, underscore the importance of mining these collections. The advent of molecular marker technology and development of mapping populations, especially AB-QTL and CSSL, have made it possible to map many of these alleles to chromosome location and begin to dissect the interactions between various genes. The fact that high quality genome sequences are now available or will soon be available, make it possible to interrogate the wild *Oryza* species accessions at a level that was not possible before. These resources will allow us to move swiftly beyond the first step of QTL identification to fine mapping traits of interest; introgressing desirable traits into elite breeding lines using markers within the gene, thus decreasing linkage drag; identifying genotype by environment interactions; determining the effect of epistasis (non-allelic genes) on traits of interest; discovering epigenetic effects such as histone modification or DNA methylation; and finally unraveling other genetic phenomenon like gene silencing. In summary, the interspecific and intergenomic mapping populations available or soon to be available, and the increased availability of SNP data, resequencing data and advanced statistical software, create even more opportunities to investigate novel alleles for agronomically important traits discovered in the *Oryza* species and increase our understanding of the mechanisms underlying these traits to deal with the challenges of climate change and feeding nine billion people.

ACKNOWLEDGEMENTS

The support of National Science Foundation-Plant Genome Project: "The Genetic Basis of Transgressive Variation in Rice" (Award no. 1026555) to Ehsan Shakiba is gratefully acknowledged. Dr. Paul L. Sanchez and Dr. Benildo G. de los Reyes are acknowledged for their critical reading of this manuscript.

REFERENCES

1. Ali ML, Sanchez P, Yu SB, Lorieux M, Eizenga GC. Chromosome segment substitution lines: a powerful tool for the introgression of valuable genes from *Oryza* wild species into cultivated rice (*O. sativa*). Rice 2010;3(4):218-34.
2. Aluko G, Martinez C, Tohme J, Castano C, Bergman C, Oard J. QTL mapping of grain quality traits from the interspecific cross *Oryza sativa* x *O. glaberrima*. Theoretical and Applied Genetics 2004;109(3):630-9.
3. Amante A, De L, Sitch L, Leung H, Mew T. Sheath blight (ShB) resistance in wild rices. International Rice Research Institute Newsletter 1990;15(3).
4. Amante-Bordeos A, Sitch L, Nelson R, Dalmacio R, Oliva N, Aswidinnoor H, et al. Transfer of bacterial blight and blast resistance from the tetraploid wild rice *Oryza minuta* to cultivated rice, *Oryza sativa*. Theoretical and Applied Genetics 1992;84(3-4):345-54.
5. Ammiraju JS, Song X, Luo M, Sisneros N, Angelova A, Kudrna D, et al. The *Oryza* BAC resource: a genus-wide and genome scale tool for exploring rice genome evolution and leveraging useful genetic diversity from wild relatives. Breeding Science 2010;60(5):536-43.
6. Andaya V, Tai T. Fine mapping of the *qCTS4* locus associated with seedling cold tolerance in rice (*Oryza sativa* L.). Molecular Breeding 2007;20(4):349-58.
7. Angeles E, Khush G, Heinrichs E. New genes for resistance to whitebacked planthopper [*Sogatella furcifera*] in rice. Crop Science 1981;21(1):47-50.

8. Angeles-Shim RB, Vinarao RB, Marathi B, Jena KK. Molecular analysis of *Oryza latifolia* Desv. (CCDD genome)-derived introgression lines and identification of value-added traits for rice (*O. sativa* L.) improvement. Journal of Heredity (provisional acceptance).
9. Balkunde S, Le HL, Lee HS, Kim DM, Kang JW, Ahn SN. Fine mapping of a QTL for the number of spikelets per panicle by using near-isogenic lines derived from an interspecific cross between *Oryza sativa* and *Oryza minuta*. Plant Breeding 2013;132(1):70-6.
10. Bandillo N, Raghavan C, Muyco PA, Sevilla MAL, Lobina IT, Dilla-Ermita CJ, et al. Multi-parent advanced generation inter-cross (MAGIC) populations in rice: progress and potential for genetics research and breeding. Rice 2013;6(1):1-15.
11. Bernacchi D, Tanksley S, Beck-Bunn T, Uhlig J, Eshed Y, Zamir D, et al. Advanced backcross QTL analysis in tomato. I. Identification of QTLs for traits of agronomic importance from *Lycopersicon hirsutum*. Theoretical and Applied Genetics 1998;97(3):381-97.
12. Bhuiyan M, Narimah M, Rahim HA, Abdullah M, Wickneswari R. Transgressive variants for red pericarp grain with high yield potential derived from *Oryza rufipogon* x *Oryza sativa*: field evaluation, screening for blast disease, QTL validation and background marker analysis for agronomic traits. Field Crops Research 2011;121(2):232-9.
13. Bimpong IK, Serraj R, Chin JH, Ramos J, Mendoza EM, Hernandez JE, et al. Identification of QTLs for drought-related traits in alien introgression lines derived from crosses of rice (*Oryza sativa* cv. IR64) x *O. glaberrima* under lowland moisture stress. Journal of Plant Biology 2011;54(4):237-50.
14. Blanco A, Gadaleta A, Simeone R. Variation for yield and quality components in durum wheat backcross inbred lines derived from ssp.*dicoccoides*. Bodenkultur 2004;54(3):163-70.
15. Blum A. Plant breeding for stress environments. BocaRaton, Forida, USA: CRC Press;1998.
16. Bocco R, Lorieux M, Seck P, Futakuchi K, Manneh B, Baimey H, et al. Agro-morphological characterization of a population of introgression lines derived from crosses between IR64 (*Oryza*

sativa Indica) and TOG5681 (*Oryza glaberrima*) for drought tolerance. Plant Science 2012;183:65-6.

17. Brar D, Khush G. Alien introgression in rice. Plant Molecular Biology 1997;35(1-2):35-47.
18. Brar D, Singh K. *Oryza*. In: Kole C, editor. Wild crop relatives: Genomic and breeding resources: Cereals. Dordrecht London, New York: Springer Heidelberg;2011. p 321-36.
19. Brar D, Toriyama K, Heong K, Hardy B. Broadening the gene pool of rice through introgression from wild species. Rice is life: Scientific perspectives for the 21st century. Proceedings of the World Rice Research Conference. Tsukuba, Japan, 4-7 November 2004. 2005:157-60.
20. Brondani C, Rangel P, Brondani R, Ferreira M. QTL mapping and introgression of yield-related traits from *Oryza glumaepatula* to cultivated rice (*Oryza sativa*) using microsatellite markers. Theoretical and Applied Genetics 2002;104(6-7):1192-203.
21. Brooker RW. Plant–plant interactions and enviromental change. New Phytologist 2006;171(2):271-84.
22. Cavanagh C, Morell M, Mackay I, Powell W. From mutations to MAGIC: resources for gene discovery, validation and delivery in crop plants. Current Opinion in Plant Biology 2008;11(2):215-21.
23. Cheema K, Bains N, Mangat G, Das A, Vikal Y, Brar D, et al. Development of high yielding IR64 x *Oryza rufipogon* (Griff.) introgression lines and identification of introgressed alien chromosome segments using SSR markers. Euphytica 2008a;160(3):401-9.
24. Cheema KK, Grewal NK, Vikal Y, Sharma R, Lore JS, Das A, et al. A novel bacterial blight resistance gene from *Oryza nivara* mapped to 38kb region on chromosome 4L and transferred to *Oryza sativa* L. Genetical Research 2008b;90(5):397.
25. Chen J, Huang DR, Wang L, Liu GJ, Zhuang JY. Identification of quantitative trait loci for resistance to whitebacked planthopper, *Sogatella furcifera*, from an interspecific cross *Oryza sativa* x *O. rufipogon*. Breeding Science 2010;60(2):153-9.
26. Chen L, Zhang W, Cheng X, Yang Y, Yan C, Wang X, et al. Identification of quantitative trait loci for bacterial blight

resistance derived from *Oryza meyeriana* and agronomic traits in recombinant inbred lines of *Oryza sativa*. Journal of Phytopathology 2012;160(9):461-8.

27. Chen Z, Hu F, Xu P, Li J, Deng X, Zhou J, et al. QTL analysis for hybrid sterility and plant height in interspecific populations derived from a wild rice relative, *Oryza longistaminata*. Breeding Science 2009;59(4):441-5.
28. Cho Y, Suh J, Choi I, Hong H, Baek M, Kang K, et al. QTLs analysis of yield and its related traits in wild rice relative *Oryza rufipogon*. Trends in Crop Research 2003;4:19-29.
29. Couch B, Kohn L. A multilocus gene genealogy concordant with host preference indicates segregation of a new species, *Magnaporthe oryzae*, from*M. grisea*. Mycologia 2002;94(4):683-93.
30. Deen R, Ramesh K, Gautam S, Rao Y, Lakshmi V, Viraktamath B, et al. Identification of new gene for BPH resistance introgressed from *O. rufipogon*. Rice Genetics Newsletter 2010;25:70-1.
31. Doi K, Taguchi A, Yoshimura A. A new locus affecting high F1 pollen sterility found in backcross progenies between *Japonica* rice and African rice. Rice Genetics Newsletter 1998;15:146-8.
32. Doi K, Taguchi K, Yoshimura A. RFLP mapping of *S20* and *S21* for F1 pollen semi-sterility found in backcross progeny of *Oryza sativa* and *O. glaberrima*. Rice Genetics Newsletter 1999;16:65-8.
33. Eizenga G, Jia Y, Agrama H, Lee F. Exploring genetic diversity and potential novel disease resistance genes in a collection of rice (*Oryza* spp.) wild relatives. Genetic Resources and Crop Evolution 2009;56(1):65-76.
34. Eizenga GC, Prasad B, Jackson AK, Jia MH. Identification of rice sheath blight and blast quantitative trait loci in two different *O. sativa/O. nivara*advanced backcross populations. Molecular Breeding 2013;31:889–907.
35. Eizenga G, Neves P, Bryant RJ, Agrama H, Mackill DJ. Evaluation of a M-202 x *Oryza nivara* advanced backcross mapping population for seedling vigor, agronomic traits, yield components, yield and grain quality. Euphytica (provisional acceptance).
36. Eshed Y, Zamir D. An introgression line population of *Lycopersicon*

pennellii in the cultivated tomato enables the identification and fine mapping of yield-associated QTL. Genetics 1995;141(3):1147.

37. Foy CD. Soil chemical factors limiting plant root growth. Limitations to plant root growth: Springer; 1992. p 97-149.
38. Fu Q, Tan L, Zhu Z, Ma D, Fu Y, Cai H, et al. Analysis of QTLs for yield-related traits in Yuanjiang common wild rice (*Oryza rufipogon* Griff.). Journal of Genetics and Genomics 2010;37(2):147-57.
39. Fu X, Lu Y, Liu X, Li J. Progress on transferring elite genes from non-AA genome wild rice into *Oryza sativa* through interspecific hybridization. Rice Science 2008;15(2):79-87.
40. Fujino K, Sekiguchi H, Sato T, Kiuchi H, Nonoue Y, Takeuchi Y, et al. Mapping of quantitative trait loci controlling low-temperature germinability in rice (*Oryza sativa* L.). Theoretical and Applied Genetics 2004;108(5):794-9.
41. Fujita D, Doi K, Yoshimura A, Yasui H. Mapping a new resistance gene for green rice leafhopper introgressed from *Oryza rufipogon* Griff. into cultivated rice, *Oryza sativa* L. Rice Genetics Newletter 2003;20:79-80.
42. Fujita D, Doi K, Yoshimura A, Yasui H. Introgression of a resistance gene for green rice leafhopper from *Oryza nivara* into cultivated rice, *Oryza sativa* L. Rice Genetics Newsletter 2004;21: 64–6.
43. Fujita D, Doi K, Yoshimura A, Yasui H. Molecular mapping of a novel gene, *Grh5*, conferring resistance to green rice leafhopper (*Nephotettix cincticeps* Uhler) in rice, *Oryza sativa* L. Theoretical and Applied Genetics 2006;113(4):567-73.
44. Fujita D, Doi K, Yoshimura A, Yasui H. A major QTL for resistance to green rice leafhopper (*Nephotettix cincticeps* Uhler) derived from African rice (*Oryza glaberrima* Steud.). Breeding Science 2010;60(4):336-41.
45. Fujita D, Yoshimura A, Yasui H. Development of near-isogenic lines and pyramided lines carrying resistance genes to green rice leafhopper (*Nephotettix cincticeps* Uhler) with the Taichung 65 genetic background in rice (*Oryza sativa* L.). Breeding Science 2010;60(1):18-27.
46. Fukuoka S, Nonoue Y, Yano M. Germplasm enhancement by developing advanced plant materials from diverse rice accessions. Breeding Science 2010;60(5):509-17.

47. Fukuta Y, Tamura K, Hirae M, Oya S. Genetic analysis of resistance to green rice leafhopper (*Nephotettix cincticeps* Uhler) in rice parental line, Norin-PL6, using RFLP markers. Breeding Science 1998;48:243–9.

48. Fuller D, Castillo C, Weisskopf A, Kingwell-Banham E, Song J, Sato Y, et al. Consilience of genetics and archaeobotany in the entangled history of rice. Archaeological and Anthropological Sciences 2010:1-17.

49. Fulton T, Beck-Bunn T, Emmatty D, Eshed Y, Lopez J, Petiard V, et al. QTL analysis of an advanced backcross of *Lycopersicon peruvianum* to the cultivated tomato and comparisons with QTLs found in other wild species. Theoretical and Applied Genetics 1997;95(5-6):881-94.

50. Futakuchi K, Fofana M, Sié M. Varietal differences in lodging resistance of African rice (*Oryza glaberrima* Steud.). Asian Journal of Plant Sciences 2008;7(6):569-73.

51. Garris A, Coburn J, Kresovich S, McCouch S, Tai T. Genetic structure and diversity in *Oryza sativa* L. Genetics 2005;169(3):1631-8.

52. Ghesquière A, Séquier J, Second G, Lorieux M. First steps towards a rational use of African rice, *Oryza glaberrima*, in rice breeding through a 'contig line'concept. Euphytica 1997;96(1):31-9.

53. Godfray HCJ, Beddington JR, Crute IR, Haddad L, Lawrence D, Muir JF, et al. Food security: the challenge of feeding 9 billion people. Science 2010;327(5967):812-8.

54. Goff SA, Ricke D, Lan TH, Presting G, Wang R, Dunn M, Glazebrook J, Sessions A, Oeller P, Varma H et al. A draft sequence of the rice genome (*Oryza sativa* L. ssp. *japonica*). Science 2002;296:92–100.

55. Grandillo S, Tanksley SD. Advanced backcross QTL analysis: Results and perspectives. In: Tuberosa R, Phillips RL, Gale M. (editors) In the Wake of the Double Helix: From the Green Revolution to the Gene Revolution, Proceedings of the International Congress, 27-31 May 2003 Bologna, Italy; 2005.

56. Grennan AK. Abiotic stress in rice. An "omic" approach. Plant Physiology 2006;140(4):1139-41.

57. Gu K, Tian D, Yang F, Wu L, Sreekala C, Wang D, et al. High-resolution genetic mapping of *Xa27(t)*, a new bacterial blight

resistance gene in rice,*Oryza sativa* L. Theoretical and Applied Genetics 2004;108:800-7.

58. Guo SB, Wei Y, Li XQ, Liu KQ, Huang FK, Chen CH, et al. Development and identification of introgression lines from cross of *Oryza sativa* and*Oryza minuta*. Rice Science 2013;20(2):95-102.
59. Gutierrez A, Carabalí S, Giraldo O, Martínez C, Correa F, Prado G, et al. Identification of a *Rice stripe necrosis virus* resistance locus and yield component QTLs using *Oryza sativa* x *O. glaberrima* introgression lines. BMC Plant Biology 2010;10(1):6.
60. Haley CS, Knott SA. A simple regression method for mapping quantitative trait loci in line crosses using flanking markers. Heredity 1992;69(4):315-24.
61. Hanamaratti N, Prashanthi S, Salimath P, Hanchinal R, Mohankumar H, Parameshwarappa K, et al. Traditional and races of rice in Karnataka: reservoirs of valuable traits. Current Science 2008;94:242–7.
62. Hechanova S. Molecular characterization of introgression and putative genes for bacterial blight resistance in wide-cross derivatives of rice (*Oryza sativa*). Food and Agriculture Organization of the United Nations. 2008. http://agris.fao.org/agris-search/search.do?f=2009/PH/PH0909.xml;PH2009000711 (accessed 24 January 2014).
63. Hernandez JE, Khush GS. Genetics of resistance to white-backed planthopper in some rice (*Oryza sativa* L.) varieties. *Oryza* 1981;18(1):44-50.
64. Hirabayashi H, Angeles E, Kaji R, Ogawa T, Brar D, Khush G. Identification of brown planthopper resistance gene derived from *O. officinalis* using molecular markers in rice. Breeding Science 1998;48(Suppl 1):82.
65. Hirabayashi H, Sato H, Nonoue Y, Kuno-Takemoto Y, Takeuchi Y, Kato H, et al. Development of introgression lines derived from *Oryza rufipogon* and *O. glumaepatula* in the genetic background of japonica cultivated rice (*O. sativa* L.) and evaluation of resistance to rice blast. Breeding Science 2010;60(5):604-12.
66. Hou Ly, Yu P, Xu Q, Yuan XP, Yu HY, Wang YP, et al. Genetic analysis and preliminary mapping of two recessive resistance

genes to brown planthopper, *Nilaparvata lugens* Stål in rice. Rice Science 2011;18(3):238-42.

67. Hu F, Xu P, Deng X, Zhou J, Li J, Tao D. Molecular mapping of a pollen killer gene *S29(t)* in *Oryza glaberrima* and co-linear analysis with *S22* in *O. glumaepatula*. Euphytica 2006;151(3):273-8.
68. Hu G, Zhang D, Pan H, Li B, Wu J, Zhou X, et al. Fine mapping of the awn gene on chromosome 4 in rice by association and linkage analyses. Chinese Science Bulletin 2011;56(9):835-9.
69. Hu H, You J, Fang Y, Zhu X, Qi Z, Xiong L. Characterization of transcription factor gene SNAC2 conferring cold and salt tolerance in rice. Plant Molecular Biology 2008;67(1-2):169-81.
70. Huang X, Kurata N, Wei X, Wang ZX, Wang A, Zhao Q, et al. A map of rice genome variation reveals the origin of cultivated rice. Nature 2012;490(7421):497-501.
71. Huang Z, He G, Shu L, Li X, Zhang Q. Identification and mapping of two brown planthopper resistance genes in rice. Theoretical and Applied Genetics 2001;102(6-7):929-34.
72. Imai I, Kimball JA, Conway B, Yeater KM, McCouch SR, McClung A. Validation of yield-enhancing quantitative trait loci from a low-yielding wild ancestor of rice. Molecular Breeding 2013;32:1-20.
73. Ingram J, Bartels D. The molecular basis of dehydration tolerance in plants. Annual Review of Plant Biology 1996;47(1):377-403.
74. Ishii T, Brar D, Multani D, Khush G. Molecular tagging of genes for brown planthopper resistance and earliness introgressed from *Oryza australiensis* into cultivated rice, *O. sativa*. Genome 1994;37(2):217-21.
75. Jacquemin J, Bhatia D, Singh K, Wing RA. The international *Oryza* map alignment project: development of a genus-wide comparative genomics platform to help solve the 9 billion-people question. Current Opinion in Plant Biology 2013;16:147-56.
76. Jena KK. The species of the genus *Oryza* and transfer of useful genes from wild species into cultivated rice, *O. sativa*. Breeding Science 2010;60(5):518-23.
77. Jena K, Jeung J, Lee J, Choi H, Brar D. High-resolution mapping of a new brown planthopper (BPH) resistance gene, *Bph18(t)*,

and marker-assisted selection for BPH resistance in rice (*Oryza sativa* L.). Theoretical and Applied Genetics 2006;112(2):288-97.

78. Jena K, Khush GS. Monosomic alien addition lines of rice: production, morphology, cytology, and breeding behavior. Genome 1989;32(3):449-55.

79. Jena KK, Khush G. Introgression of genes from *Oryza officinalis* Well Ex. Watt to cultivated rice, *O. sativa* L. Theoretical and Applied Genetics 1990;80:737-45.

80. Jena K, Kochert G, Khush G. RFLP analysis of rice (*Oryza sativa* L.) introgression lines. Theoretical and Applied Genetics 1992;84(5-6):608-16.

81. Jena K, Pasalu I, Rao Y, Varalaxmi Y, Krishnaiah K, Khush G, et al. Molecular tagging of a gene for resistance to brown planthopper in rice (*Oryza sativa* L.). Euphytica 2002;129(1):81-8.

82. Jeung JU, Kim BR, Cho YC, Han SS, Moon HP, Lee YT, Jena KK. A novel gene, *Pi40(t)*, linked to the DNA markers derived from NBS-LRR motifs confers broad spectrum of blast resistance in rice. Theoretical and Applied Genetics 2007;115(8):1163-77.

83. Jia Y, Liu G, Costanzo S, Lee S, Dai Y. Current progress on genetic interactions of rice with rice blast and sheath blight fungi. Frontiers of*Agriculture* in *China* 2009;3(3):231–9.

84. Jin F, Kwon S, Kang K, Jeong O, Le L, Yoon D, et al. Introgression for grain traits from *Oryza minuta* into rice, *O. sativa*. Rice Genetics Newsletter 2004; 21:15.

85. Jin FX, Ji SD, Xie XB, Kang JW, Ju HG, Ahn SN. Detection of epistatic interaction of two QTLs, *gw8.1* and *gw9.1*, underlying grain weight using nearly isogenic lines in rice. Breeding Science 2011;61(1):69-75.

86. Jin FX, Kim DM, Ju HG, Ahn SN. Mapping quantitative trait loci for awnness and yield component traits in isogenic lines derived from an *Oryza sativa/O. rufipogon* cross. Journal of Crop Science and Biotechnology 2009;12(1):9-15.

87. Joehanes R, Nelson JC. QGene 4.0, an extensible Java QTL-analysis platform. Bioinformatics 2008;24(23):2788-9.

88. Jones DB, Peterson ML. Rice seedling vigor at sub-optimal temperatures. Crop Science 1976;16(1):102-5.

89. Kang J, Suh J, Kim D, Oh C, Oh J, Ahn S. QTL mapping of

agronomic traits in an advanced backcross population from a cross between *Oryza sativa*L. cv. Milyang 23 and *O. glaberrima*. Korean Journal of Breeding Science 2008;40(3):243-9.

90. Kao CH, Zeng ZB, Teasdale RD. Multiple interval mapping for quantitative trait loci. Genetics 1999;152(3):1203-16.
91. Kellogg EA. The evolutionary history of Ehrhartoideae, Oryzeae, and *Oryza*. Rice 2009;2(1):1-14.
92. Keurentjes JJB, Bentsink L, Hanhart CJ, Vries HBD, Koornneef M, Vreugdenhil D, et al. Development of a near-isogenic line population of*Arabidopsis thaliana* and comparison of mapping power with a recombinant inbred line population. Genetics 2007;175(2):891-905.
93. Khush GS. Disease and insect resistance in rice. Advances in Agronomy 1977;29:265-341.
94. Khush G, Ling K, Aquino R, Aguiero V. Breeding for resistance to grassy stunt in rice. Plant Breeding Papers; International Congress of the Society for the Advancement of Breeding Researches in Asia & Oceania. 1977.
95. Khush GS, Bacalangco E, Ogawa T. A new gene for resistance to bacterial blight from *O. longistaminata*. Rice Genetic Newsletter 1990;7:121-2.
96. Koide Y, Shinya Y, Ikenaga M, Sawamura N, Matsubara K, Onishi K, et al. Complex genetic nature of sex-independent transmission ratio distortion in Asian rice species: the involvement of unlinked modifiers and sex-specific mechanisms. Heredity 2012;108(3):242-7.
97. Koseki M, Kitazawa N, Yonebayashi S, Maehara Y, Wang ZX, Minobe Y. Identification and fine mapping of a major quantitative trait locus originating from wild rice, controlling cold tolerance at the seedling stage. Molecular Genetics and Genomics 2010;284(1):45-54.
98. Kover PX, Scarcelli N, Valdar W, Trakalo J, Durrant C, Mott R, et al. A multiparent advanced generation inter-cross to fine-map quantitative traits in *Arabidopsis thaliana*. PLoS Genetics 2009;5(7). DOI: 10.1371/journal.pgen.1000551
99. Kurakazu T, Sobrizal, Yoshimura A. RFLP mapping of genes for awn on chromosomes 4 and 5 in rice using *Oryza*

meridionalis introgression lines. Rice Genetics Newsletter 2001(18):28-30.

100. Lakshmanan P, Velusamy R. Resistance to sheath blight (ShB) and brown spot (BS) in lines derived from *Oryza officinalis*. International Rice Resesarch Institute Newsletter 1991;16:8.
101. Lander ES, Botstein D. Mapping Mendelian factors underlying quantitative traits using RFLP linkage maps. Genetics 1989;121(1):185-99.
102. Lee FN, Rush MC. Rice sheath blight: a major rice disease. Plant Disease 1983;67:829–32.
103. Lee H, Sasaki K, Higashitani A, Ahn S, Sato T. Mapping and characterization of quantitative trait loci for mesocotyl elongation in rice (*Oryza sativa* L.). Rice 2012;5(13).
104. Lee SJ, Oh CS, Suh JP, McCouch S, Ahn SN. Identification of QTLs for domestication-related and agronomic traits in an *Oryza sativa* x *O. rufipogon* BC1F7 population. Plant Breeding 2005;124(3):209-219.
105. Lei D, Tan L, Liu F, Chen L, Sun C. Identification of heat-sensitive QTL derived from common wild rice (*Oryza rufipogon*, Griff.). Plant Science 2013;201-202:123-7.
106. Lexer C, Fay M. Adaptation to environmental stress: a rare or frequent driver of speciation? Journal of Evolutionary Biology 2005;18(4):893-900.
107. Li C, Zhou A, Sang T. Genetic analysis of rice domestication syndrome with the wild annual species, *Oryza nivara*. New Phytologist 2006;170(1):185-94.
108. Li J, Thomson M, McCouch SR. Fine mapping of a grain-weight quantitative trait locus in the pericentromeric region of rice chromosome 3. Genetics 2004;168(4):2187-95.
109. Li J, Xiao J, Grandillo S, Jiang L, Wan Y, Deng Q, et al. QTL detection for rice grain quality traits using an interspecific backcross population derived from cultivated Asian (*O. sativa* L.) and African (*O. glaberrima* S.) rice. Genome 2004;47(4):697-704.
110. Li JB, Xia MY, Qi HX, He GC, Wan BL, Zha ZP. Marker assisted selection for brown planthopper (*Nilaparvata lugens* Stål)

resistance genes *Bph14*and *Bph15* in rice. Scientia Agricultura Sinica 2006;39(10):2132-7.

111. Li Z, Pinson SR, Stansel JW, Paterson AH. Genetic dissection of the source-sink relationship affecting fecundity and yield in rice (*Oryza sativa* L.). Molecular Breeding 1998;4(5):419-26.

112. Li ZX, Septiningsih EM, Quilloy-Mercado SM, McNally KL, Mackill DJ. Identification of *SUB1A* alleles from wild rice *Oryza rufipogon* Griff. Genetic Resources and Crop Evolution 2011;58(8):1237-42.

113. Linh LH, Hang NT, Jin FX, Kang KH, Lee YT, Kwon SJ, et al. Introgression of a quantitative trait locus for spikelets per panicle from *Oryza minuta*to the *O. sativa* cultivar Hwaseongbyeo. Plant Breeding 2008;127(3):262-7.

114. Linh LH, Jin FX, Kang KH, Lee YT, Kwon SJ, Ahn SN. Mapping quantitative trait loci for heading date and awn length using an advanced backcross line from a cross between *Oryza sativa* and *O. minuta*. Breeding Science 2006;56(4):341-9.

115. Liu G, Lu G, Zeng L, Wang GL. Two broad-spectrum blast resistance genes, *Pi9(t)* and *Pi2(t)*, are physically linked on rice chromosome 6. Molecular Genetics and Genomics. 2002;267(4):472-80.

116. Liu G, Yan H, Fu Q, Qian Q, Zhang Z, Zhai W, et al. Mapping of a new gene for brown planthopper resistance in cultivated rice introgressed from*Oryza eichingeri*. Chinese Science Bulletin 2001;46(17):1459-62.

117. Liu G, Yang J, Xu H, Hayat Y, Zhu J. Genetic analysis of grain yield conditioned on its component traits in rice (*Oryza sativa* L.). Crop and Pasture Science 2008;59(2):189-95.

118. Lorieux M. MapDisto: Fast and efficient computation of genetic linkage maps. Molecular Breeding 2012;30(2):1231-5.

119. Lu BR, Jackson M, Vaughn D. Wild rice taxonomy. In: Rice knowledgebank, International Rice Research Institute, Philippines. 2014; http://www.knowledgebank.irri.org/extension/index.php/wild-rice-taxonomy (accessed 23 January 2014).

120. Luo DG, Zhang SP. Preliminary study on simplified distant cross techniques by using male-sterile lines as maternal parent and *O. gradiglumis* as paternal parent. Seed 2000;3:49–50.

121. Manneh B, Kiepe P, Sie M, Ndjiondjop M, Drame N, Traore K, et al. Exploiting partnerships in research and development to help African rice farmers cope with climate variability. Journal of SAT Agricultural Research 2007;4(1):1-10.

122. Marri PR, Sarla N, Reddy LV, Siddiq E. Identification and mapping of yield and yield related QTLs from an Indian accession of *Oryza rufipogon*. BMC Genetics 2005;6(1):33.

123. Matsushita S, Kurakazu T, Sobrizal, Doi K, Yoshimura A. Mapping genes for awn in rice using *Oryza meridionalis* introgression lines. Rice Genetics Newsletter 2003;20:17.

124. Matsushita S, Sanchez P, Sobrizal, Doi K, Yoshimura A. Identification of new alleles of awnness genes, *An7* and *An8*, in rice using *Oryza glumaepatula* introgression lines. Rice Genetics Newsletter 2003;20:19-20.

125. McCombie R. New whole genome de-novo assemblies of three divergent strains of rice document novel gene space in *aus* and *indica* missed by previous reference assemblies. 7th International Rice Genetics Symposium Program and Abstract Book; Manila, Philippines: 5-8 November 2013; p 120.

126. McCouch SR, McNally KL, Wang W, Hamilton RS. Genomics of gene banks: a case study in rice. American Journal of Botany 2012;99(2):407–23.

127. McCouch SR, Sweeney M, Jiang H, Thomson M, Septiningsih E, Edwards J, et al. Through the genetic bottleneck: *O. rufipogon* as a source of trait-enhancing alleles for *O. sativa*. Euphytica 2007;154(3):317-39.

128. McCouch SR, Zhao K, Wright M, Tung CW, Ebana K, Thomson M, et al. Development of genome-wide SNP assays for rice. Breeding Science 2010;60(5):524-35.

129. Mew T, Alvarez A, Leach J, Swings J. Focus on bacterial blight of rice. Plant Disease 1993;77(1):5-12.

130. Miura K, Yamamoto E, Morinaka Y, Takashi T, Kitano H, Matsuoka M, et al. The hybrid breakdown 1 (t) locus induces interspecific hybrid breakdown between rice *Oryza sativa* cv. Koshihikari and its wild relative *O. nivara*. Breeding Science 2008;58(2):99-105.

131. Moncada P, Martinez C, Borrero J, Châtel M, Gauch Jr H, Guimaraes E, et al. Quantitative trait loci for yield and yield

components in an *Oryza sativa* x *Oryza rufipogon* BC2F2 population evaluated in an upland environment. Theoretical and Applied Genetics 2001;102(1):41-52.

132. Multani D, Jena K, Brar D, de los Reyes B, Angeles E, Khush G. Development of monosomic alien addition lines and introgression of genes from*Oryza australiensis* Domin. to cultivated rice *O. sativa* L. Theoretical and Applied Genetics 1994;88(1):102-109.
133. Multani D, Khush G, de los Reyes B, Brar D. Alien genes introgression and development of monosomic alien addition lines from *Oryza latifolia*Desv. to rice, *Oryza sativa* L. Theoretical and Applied Genetics 2003;107(3):395-405.
134. Hanamaratti NG, Prashanthi SK, Salimath PM, Hanchinal RR, Mohankumar HD, Parameshwarappa KG, Raikar SD. Traditional land races of rice in Karnataka: reservoirs of valuable traits. Current Science 2008;94:242-7.
135. Nagao S, Takahashi T. Trial construction of twelve linkage groups in Japanese rice. Genetical studies on rice plant, XXVII Journal of the Faculty of Agriculture, Hokkaido University. 1963:72-130.
136. Ndjiondjop MN, Manneh B, Cissoko M, Drame NK, Kakai RG, Bocco R, et al. Drought resistance in an interspecific backcross population of rice (*Oryza* spp.) derived from the cross WAB56-104 (*O. sativa*) x CG14 (*O. glaberrima*). Plant Science 2010;179(4):364-73.
137. Ndjiondjop MN, Manneh B, Cissoko M, Drame NK, Kakai RG, Bocco R, et al. Drought resistance in an interspecific backcross population of rice (*Oryza* spp.) derived from the cross WAB56-104 (*O. sativa*) x CG14 (*O. glaberrima*). Plant Science 2010;179(4):364-73.
138. Ndjiondjop MN, Albar L, Fargette D, Fauquet C, Ghesquière A. The genetic basis of high resistance to rice yellow mottle virus (RYMV) in cultivars of two cultivated rice species. Plant Disease 1999;83(10):931-35.
139. Nguyen BD, Brar DS, Bui BC, Nguyen TV, Pham LN, Nguyen HT. Identification and mapping of the QTL for aluminum tolerance introgressed from the new source, *Oryza rufipogon* Griff., into indica rice (*Oryza sativa* L.). Theoretical and Applied Genetics 2003;106(4):583-93.

140. Ni J, Colowit P, Oster J, Xu K, Mackill D. Molecular markers linked to stem rot resistance in rice. Theoretical and Applied Genetics 2001;102(4):511-6.

141. Niño-Liu DO, Ronald PC, Bogdanove AJ. *Xanthomonas oryzae* pathovars: model pathogens of a model crop. Molecular Plant Pathology 2006;7(5):303-24.

142. Nuque F, Aguiero V, Ou S. Inheritance of resistance to grassy stunt virus in rice. Plant Disease 1982;66(1):63-4.

143. Oh JM, Balkunde S, Yang P, Yoon DB, Ahn SN. Fine mapping of grain weight QTL, *tgw11* using near isogenic lines from a cross between *Oryza sativa* and *O. grandiglumis*. Genes and Genomics 2011;33(3):259-65.

144. Orjuela J, Garavito A, Bouniol M, Arbelaez JD, Moreno L, Tohme J, et al. A universal core genetic map for rice. Theoretical and Applied Genetics 2010;120(3):563-72.

145. Oryzabase. List of Gene 2014 (Updated 20 Jan. 2014) http://www.shigen.nig.ac.jp/rice/oryzabaseV4/gene/list (accessed 24 January 2014).

146. Ou SH. Rice Diseases. Commonwealth Agricultural Bureau, Kew, UK. 1985.

147. Pastori GM, Foyer CH. Common components, networks, and pathways of cross-tolerance to stress. The central role of "redox" and abscisic acid-mediated controls. Plant Physiology 2002;129(2):460-8.

148. Pinson SRM, Capdevielle FM, Oard JH. Confirming QTLs and finding additional loci conditioning sheath blight resistance in rice using recombinant inbred lines. Crop Science 2005; 45:503–10

149. Plowright RA, Nash P, Coyne DL, Jones MP. Resistance to the rice nematodes *Heterodera sacchari*, Meloidogyne graminicola and *M. incognita* in*Oryza glaberrima* and *O. glaberrima* x *O. sativa* interspecific hybrids. Nematology 1999;1(7-8):745-51.

150. Poehlman JM, Sleper DA. Breeding Field Crops, Fourth edition: Iowa State University Press 1995.

151. Prasad B, Eizenga GC. Rice sheath blight disease resistance identified in *Oryza* spp. accessions. Plant Disease 2008;92(11):1503-9.

152. Qu S, Liu G, Zhou B, Bellizzi M, Zeng L, Wang GL, et al. The broad-spectrum blast resistance gene *Pi9* encodes a nucleotide-binding site-leucine-rich repeat protein and is a member of a multigene family in rice. Genetics 2006;172(3):1901-14.

153. Wing R. 15 million years of *Oryza* evolution as seen through the analysis of 12 high quality genome assemblies. 7th International Rice Genetics Symposium Program and Abstract Book; Manila, Philippines: 5-8 November 2013; p 121-2.

154. Rahman ML, Jiang W, Chu SH, Qiao Y, Ham TH, Woo MO, Lee J, Khanam MS, Chin JH, Jeung JU, Brar DS, Jena KK, Koh HJ. High-resolution mapping of two rice brown planthopper resistance genes, *Bph20 (t)* and *Bph21 (t)*, originating from *Oryza minuta*. Theoretical and Applied Genetics 2009;119(7):1237-46.

155. Rahman L, Koh HJ, Khanam MS. QTL analysis for yield related traits using populations derived from an indica-japonica hybrid in rice (*Oryza sativa* L.). Czech Journal of Genetics and Plant Breeding 2008;44(3):93-104.

156. Ram T, Deen R, Gautam S, Ramesh K, Rao Y, Brar D. Identification of new genes for brown planthopper resistance in rice introgressed from *O. glaberrima* and *O. minuta*. Rice Genetics Newsletter 2010;25:67.

157. Ram T, Laha G, Gautam S, Deen R, Madhav MS, Brar D, et al. Identification of new gene introgressed from *Oryza brachyantha* with broad-spectrum resistance to bacterial blight of rice in India. Rice Genetics Newsletter. 2010;25:57-8.

158. Rangel PN, Brondani RPV, Rangel PHN, Brondani C. Agronomic and molecular characterization of introgression lines from the interspecific cross*Oryza sativa* (BG 90-2) x *Oryza glumaepatula* (RS-16). Genetic and Molecular Research 2008;7(1):184-95.

159. Rangel PN, Vianello RP, Melo ATO, Rangel PHN, Mendonça JA, Brondani C. Yield QTL analysis of *Oryza sativa* x *O. glumaepatula* introgression lines. Pesquisa Agropecuária Brasileira 2013;48(3):280-6.

160. Rausher MD. Co-evolution and plant resistance to natural enemies. Nature 2001;411(6839):857-64.

161. Ravinder SS, Khush GS, Heinrichs EA. Genetic analysis of resistance

to whitebacked planthopper, *Sogatella furcifera* (Horvath), in some rice varieties. Crop Protection 1982;1:289-97.

162. Renganayaki K, Fritz AK, Sadasivam S, Pammi S, Harrington SE, McCouch SR, et al. Mapping and progress toward map-based cloning of brown planthopper biotype-4 resistance gene introgressed from *Oryza officinalis* into cultivated rice. Crop Science 2002;42(6):2112-2117.
163. Ronald PC, Albano B, Tabien R, Abenes L, Wu KS, McCouch S, et al. Genetic and physical analysis of the rice bacterial blight disease resistance locus, *Xa21*. Molecular and General Genetics 1992;236(1):113-20.
164. Saka N, Toyama T, Tsuji T, Nakamae H, Izawa T. Fine mapping of green ricehopper resistant gene *Grh-3(t)* and screening of *Grh-3(t)* among green ricehopper resistant and green leafhopper resistant cultivars in rice. Breeding Science 1997; 47(1):55.
165. Sanchez PL, Sobrizal, Ikeda K, Yasui H, Yoshimura A. Identifying late heading genes in rice using *Oryza glumaepatula* introgression lines. In: Advances in Rice Genetics. International Rice Research Institute, Los Baños, Laguna, Philippines. 2003; p 153-4.
166. Sanchez PL, Wing RA, Brar DS. The wild relative of rice: genomes and genomics. In: Zhang Q, Wing RA, editors. Genetics and Genomics of Rice, Plant Genetics and Genomics: Crops and Models 5. New York: Springer Science: 2013; p 9-25.
167. Sano Y. A new gene controlling sterility in F1 hybrids of two cultivated rice species: Its association with photoperiod sensitivity. Journal of Heredity 1983;74(6):435-9.
168. Sano Y. Sterility barriers between *Oryza sativa* and *O. glaberrima*. In: Rice genetics, Proceedings of the International Rice Genetics Symposium, International Rice Research Institute, Los Baños, Philippines: 27–31 May 1985; p 109–18.
169. Sano Y. The genic nature of gamete eliminator in rice. Genetics 1990;125(1):183-91.
170. Sarla N, Swamy BM. *Oryza glaberrima*: a source for the improvement of *Oryza sativa*. Current Science 2005;89(6):955-63.
171. Savary S, Willocquet L, Elazegui FA, Castilla NP, Teng PS. Rice pest constraints in tropical Asia: quantification of yield losses due

to rice pests in a range of production situations. Plant Disease 2000;84(3):357-69.

172. Septiningsih E, Prasetiyono J, Lubis E, Tai T, Tjubaryat T, Moeljopawiro S, et al. Identification of quantitative trait loci for yield and yield components in an advanced backcross population derived from the *Oryza sativa* variety IR64 and the wild relative *O. rufipogon*. Theoretical and Applied Genetics 2003;107(8):1419-32.
173. Shan JX, Zhu MZ, Shi M, Gao JP, Lin HX. Fine mapping and candidate gene analysis of *spd6*, responsible for small panicle and dwarfness in wild rice (*Oryza rufipogon* Griff.). Theoretical and Applied Genetics 2009;119(5):827-36.
174. Shim RA, Ashikari M, Angeles ER, Takashi T. Development and evaluation of *Oryza glaberrima* Steud. chromosome segment substitution lines (CSSLs) in the background of *O. sativa* L. cv. Koshihikari. Breeding Science 2010;60(5):613-9.
175. Sidhu G, Khush G. Linkage relationships of some genes for disease and insect resistance and semidwarf stature in rice. Euphytica 1979;28(2):233-7.
176. Sié M, Blaise K, Drissa S, Yonnelle D. Selection of intraspecific (*Oryza sativa* x *O. sativa*) and inter-specific (*O. sativa* x *O. glaberrima*) lines for their tolerance to blast in Burkina Faso. Institut de l'Environnement et de Recherches Agricoles (INERA) 01 BP 910 Bobo-Dioulasso 1:23-37.
177. Silué D, Nottéghem JL, Chaume J. Etude de la résistance spécifique d'*Oryza glaberrima* Steud à la pyriculariose du riz (translation Study of the specific resistance of *Oryza glaberrima* Steud to rice blast). Agronomie Tropicale 1992;42:121-9.
178. Sobrizal, Ikeda K, Sanchez PL, Yasui H, Yoshimura A. Development and evaluation *Oryza glumaepatula* introgression lines in rice, *Oryza sativa* L. Rice Genetics Newsletter 1999;16:107-8.
179. Sobrizal, Matsuzaki Y, Sanchez PL, Ikeda K, Yoshimura A. Identification of a gene for male gamete abortion in backcross progeny of *Oryza sativa*L. and *Oryza glumaepatula* Steud. Rice Genetics Newsletter 2000a;17:59-61.
180. Sobrizal, Matsuzaki Y, Sanchez PL, Ikeda K, Yoshimura A. Mapping of F1 pollen semi-sterility found in backcross progeny

of *Oryza sativa* L. and*Oryza glumaepatula* Steud. Rice Genetics Newsletter 2000b;17:61-3.

181. Suh JP, Ahn SN, Cho YC, Kang KH, Choi IS, Kim YG, Suh HS, Hwang HG. Mapping of QTLs for yield traits using an advanced backcross population from a cross between *Oryza sativa* and *O. glaberrima*. Korean Journal of Breeding Science. 2005;37(4):214-20.

182. Swamy BM, Kaladhar K, Rani NS, Prasad G, Viraktamath B, Reddy GA, et al. QTL analysis for grain quality traits in 2 BC2F2 populations derived from crosses between *Oryza sativa* cv Swarna and 2 accessions of *O. nivara*. Journal of Heredity 2012;103(3):442-52.

183. Takamure I, Kinoshita T. Linkage analysis in chromosomes 3 and 6. Rice Genetics Newsletter 1991;8:98-100.

184. TalamèV, Sanguineti M, Chiapparino E, Bahri H, Salem M, Forster B, et al. Identification of *Hordeum spontaneum* QTL alleles improving field performance of barley grown under rainfed conditions. Annals of Applied Biology 2004;144(3):309-319.

185. Tamura K, Fukuta Y., Hirae M, Oya S, Ashikawa I, Yagi.T. Mapping of the *Grh1* locus for green rice leafhopper resistance in rice using RFLP markers. Breeding Science 1999; 49:11–14.

186. Tamura K, Fukuta Y, Hirae M, Oya S, Ashikawa I., Yagi T. RFLP mapping of a new resistance gene for green rice leafhopper in Kanto PL10. Rice Genetics Newsletter 2004;21: 62–4.

187. Tan G, Weng Q, Ren X, Huang Z, Zhu L, He G. Two whitebacked planthopper resistance genes in rice share the same loci with those for brown planthopper resistance. Heredity 2004;92(3):212-7.

188. Tan GX, Ren X, Weng QM, Shi ZY, Zhu LL, He GC. Mapping of a new resistance gene to bacterial blight in rice line introgressed from *Oryza officinalis*. Acta Genetica Sinica 2004;31(7):724-9.

189. Tan L, Liu F, Xue W, Wang G, Ye S, Zhu Z, et al. Development of *Oryza rufipogon* and *O. sativa* introgression lines and assessment for yield-related quantitative trait loci. Journal of Integrative Plant Biology 2007;49(6):871-84.

190. Tan L, Zhang P, Liu F, Wang G, Ye S, Zhu Z, et al. Quantitative trait loci underlying domestication-and yield-related traits

in an *Oryza sativa* x*Oryza rufipogon* advanced backcross population. Genome 2008;51(9):692-704.

191. Tan LB, Fu YC, Liu FX, Wang XK, Sun CQ, Zhang PJ. Identification of quantitative trait loci controlling plant height and days to heading from Yuanjiang common wild rice (*Oryza rufipogon* Griff.). Acta Genetica Sinica 2004;31(10):1123-8.

192. Tanksley S, Grandillo S, Fulton T, Zamir D, Eshed Y, Petiard V, et al. Advanced backcross QTL analysis in a cross between an elite processing line of tomato and its wild relative *L. pimpinellifolium*. Theoretical and Applied Genetics 1996;92(2):213-24.

193. Tanksley S, Nelson J. Advanced backcross QTL analysis: a method for the simultaneous discovery and transfer of valuable QTLs from unadapted germplasm into elite breeding lines. Theoretical and Applied Genetics 1996;92(2):191-203.

194. Tanksley SD, McCouch SR. Seed banks and molecular maps: unlocking genetic potential from the wild. Science 1997;277(5329):1063-6.

195. Tanksley SD, Medina-Filho H, Rick CM. Use of naturally-occurring enzyme variation to detect and map genes controlling quantitative traits in an interspecific backcross of tomato. Heredity 1982;49 (1):11-25.

196. Thanh PT, Phan PDT, Mori N, Ishikawa R, Ishii T. Development of backcross recombinant inbred lines between *Oryza sativa* Nipponbare and *O. rufipogon* and QTL detection on drought tolerance. Breeding Science 2011;61(1):76-9.

197. Thiémélé D, Boisnard A, Chéron S, Ghesquière A, Albar L, Aké S, et al. Identification of a second major resistance gene to rice yellow mottle virus,*RYMV2*, in the African cultivated rice species, *O. glaberrima*. Theoretical and Applied Genetics 2010;121(1):169-79.

198. Thomson MJ, Tai TH, McClung AM, Lai XH, Hinga ME, Lobos KB, Xu Y, Martinez CP, McCouch SR. Mapping quantitative trait loci for yield, yield components and morphological traits in an advanced backcross population between *Oryza rufipogon* and the *Oryza sativa* cultivar Jefferson. Theoretical and Applied Genetics 2003;107(3):479-93.

199. Tian F, Zhu Z, Zhang B, Tan L, Fu Y, Wang X, et al. Fine mapping of a quantitative trait locus for grain number per panicle from wild rice (*Oryza rufipogon* Griff.). Theoretical and Applied Genetics 2006;113(4):619-29.

200. Tian L, Tan L, Liu F, Cai H, Sun C. Identification of quantitative trait loci associated with salt tolerance at seedling stage from *Oryza rufipogon*. Journal of Genetics and Genomics 2011;38(12):593-601.

201. Tilman D, Lehman C. Human-caused environmental change: impacts on plant diversity and evolution. Proceedings of the National Academy of Sciences 2001;98(10):5433-40.

202. Tseng S, Oster J. Registration of 87-Y-550, a rice germplasm line resistant to stem rot disease. Crop Science 1994;34(1)314.

203. Tung CW, Zhao K, Wright MH, Ali ML, Jung J, Kimball J, et al. Development of a research platform for dissecting phenotype–genotype associations in rice (*Oryza* spp.). Rice 2010;3(4):205-17.

204. Van Ooijen J. JoinMap 4. Software for the calculation of genetic linkage maps in experimental populations Kyazma BV, Wageningen, Netherlands: 2006.

205. Vaughan DA. The wild relatives of rice: a genetic resources handbook: Manila, Philippines: International Rice Research Institute;1994.

206. Vaughan DA, Ge S, Kaga A, Tomooka N. Phylogeny and biogeography of the genus *Oryza*. Rice Biology in the Genomics Era. Berlin-Heidelberg: Springer; 2008; p 219-34.

207. Vaughan DA, Morishima H. Biosystematics of the genus *Oryza* In: Smith CW, Dilday RH, editors. Rice: Origin, History, Technology, and Production 6: John Wiley & Sons; 2002; p 27-65.

208. Wang J, Wan X, Li H, Pfeiffer WH, Crouch J, Wan J. Application of identified QTL-marker associations in rice quality improvement through a design-breeding approach. Theoretical and Applied Genetics 2007;115(1):87-100.

209. Wang S, Basten CJ, Zeng ZB. Windows QTL Cartographer 2.5. Department of Statistics, North Carolina State University, Raleigh, NC.2012. http://statgen.ncsu.edu/qtlcart/WQTLCart.htm (accessed 24 January 2014).

210. Wang YM, Dong ZY, Zhang ZJ, Lin XY, Shen Y, Zhou D, et al. Extensive *de novo* genomic variation in rice induced by introgression from wild rice (*Zizania latifolia* Griseb.). Genetics 2005;170(4):1945-56.

211. Wickneswari R, Bhuiyan M, Lim LS, Thomson MJ, Narimah MK, Abdullah MZ. Identification and validation of quantitative trait loci for agronomic traits in advanced backcross breeding lines derived from *Oryza rufipogon* x *Oryza sativa* cultivar MR219. Plant Molecular Biology Reporter 2012;30(4):929-39.

212. Win KT, Kubo T, Miyazaki Y, Doi K, Yamagata Y, Yoshimura A. Identification of two loci causing F1 pollen sterility in inter-and intraspecific crosses of rice. Breeding Science 2009;59(4):411-8.

213. Win KT, Yamagata Y, Miyazaki Y, Doi K, Yasui H, Yoshimura A. Independent evolution of a new allele of F1 pollen sterility gene *S27* encoding mitochondrial ribosomal protein L27 in *Oryza nivara*. Theoretical and Applied Genetics 2011;122(2):385-94.

214. Wu P, Zhao B, Yan J, Luo A, Wu Y, Senadihra D. Genetic control of seedling tolerance to aluminum toxicity in rice. Euphytica 1997;97:289-293.

215. Xia R, Xiao N, Hong Y, Zhang C, Su Y, Zhang X, et al. QTLs mapping for cold tolerance at seedling stage in Dongxiang wild rice (*Oryza rufipogon*Griff.). Scientia Agricultura Sinica 2010;43:443-51.

216. Xiao J, Grandillo S, Ahn SN, McCouch SR, Tanksley SD, Li J, et al. Genes from wild rice improve yield. Nature (London) 1996;384(6606):223-4.

217. Xiao J, Li J, Grandillo S, Ahn SN, Yuan L, Tanksley SD, et al. Identification of trait-improving quantitative trait loci alleles from a wild rice relative, *Oryza rufipogon*. Genetics 1998;150(2):899-909.

218. Xie X, Song MH, Jin F, Ahn SN, Suh JP, Hwang HG, et al. Fine mapping of a grain weight quantitative trait locus on rice chromosome 8 using near-isogenic lines derived from a cross between *Oryza sativa* and *Oryza rufipogon*. Theoretical and Applied Genetics 2006;113(5):885-94.

219. Yan CQ, Qian KX, Yan QS, Zhang XQ, Xue GP, Huangfu WG, et al. Use of asymmetric somatic hybridization for transfer of the

bacterial blight resistance trait from *Oryza meyeriana* L. to *O. sativa* L. ssp. *japonica*. Plant Cell Reports 2004;22(8):569-75.

220. Yang J, Hu C, Hu H, Yu R, Xia Z, Ye X, et al. QTL Network: mapping and visualizing genetic architecture of complex traits in experimental populations. Bioinformatics 2008;24(5):721-3.
221. Yang H, Ren X, Weng Q, Zhu L, He G. Molecular mapping and genetic analysis of a rice brown planthopper (*Nilaparvata lugens* Stål) resistance gene. Hereditas 2002;136(1):39-43.
222. Yang H, You A, Yang Z, Zhang F, He R, Zhu L, He G. High-resolution genetic mapping at the *Bph15* locus for brown planthopper resistance in rice (*Oryza sativa* L.). Theoretical and Applied Genetics 2004;110(1):182-91.
223. Yasui H, Iwata N. Production of monosomic alien addition lines of *Oryza sativa* having a single *O. punctata* chromosome. In: Rice Genetics II: Proceedings of the 2nd International Rice Genetics Symposium. 14-18 May 1990. International Rice Research Institute;1991; p 147-55.
224. Yasui H, Yoshimura A. QTL mapping of antibiosis to green leafhopper, *Nephotettix virescens* Distant and green rice leafhopper, *Nephotettix cincticeps* Uhler in rice, *Oryza sativa* L. Rice Genetics Newsletter 1999;16:96-8.
225. Yazawa S, Yasui H, Yoshimura A, Iwata N. RFLP mapping of genes for resistance to green rice leafhopper (*Nephotettix cincticeps* Uhler) in rice cultivar DV85 using near isogenic lines. Science Bulletin of the Faculty of Agriculture, Kyushu University 1988; 52:169–75.
226. Yoon DB, Kang KH, Kim HJ, Ju HG, Kwon SJ, Suh JP, et al. Mapping quantitative trait loci for yield components and morphological traits in an advanced backcross population between *Oryza grandiglumis* and the *O. sativa* japonica cultivar Hwaseongbyeo. Theoretical and Applied Genetics 2006;112(6):1052-62.
227. Yu J, Hu S, Wang J, Wong GKS, Li S, Liu B, et al. A draft sequence of the rice genome (*Oryza sativa* L. ssp. *indica*). Science 2002;296(5565):79-92.
228. Yuan PR, Kim HJ, Chen QH, Ju HG, Lee SJ, Ji SD, et al. QTL dissection of agronomic and domestication traits using introgression lines carrying wild rice (*Oryza rufipogon* Griff.)

segments in cultivated rice (*O. sativa* L.) background. Journal of Crop Science and Biotechnology 2009;12(4):245-52.

229. Zeng ZB. Precision mapping of quantitative trait loci. Genetics 1994;136:1457-68.
230. Zhang Q, Lin SC, Zho BY, Wang CL, Yang WC, Zhou YL, Li DY, Chen CB,. Zhu LH. Identification and tagging a new gene for resistance to bacterial blight (*Xanthomonas oryzae* pv. oryzae) from *O. rufipogon*. Rice Genetics Newsletter 1998;15: 138-42.
231. Zhang Q. Strategies for developing green super rice. Proceedings of the National Academy of Sciences (USA) 2007;104(42):16402-9.
232. Zhang X, Zhou S, Fu Y, Su Z, Wang X, Sun C. Identification of a drought tolerant introgression line derived from Dongxiang common wild rice (*O. rufipogon* Griff.). Plant Molecular Biology 2006;62(1-2):247-59.
233. Zhao K, Tung CW, Eizenga GC, Wright MH, Ali ML, Price AH, Norton GJ, Islam MR, Reynolds A, Mezey J, McClung AM, Bustamante CD, McCouch SR. Genome-wide association mapping reveals a rich genetic architecture of complex traits in *Oryza sativa*. Nature Communications 2011;2:467. DOI: 10.1038/ncomms1467
234. Zhou F, He H, Chen H, Yu H, Lorieux M, He Y. Genomics-based breeding technology.In: Zhang Q, Wing RA, editors. Genetics and Genomics of Rice, Plant Genetics and Genomics: Crops and models 5. New York: Springer; 2013; p 329-48.

Chapter 9

Genes and QTLs Resistant to Biotic and Abiotic Stresses from Wild Rice and Their Applications in Cultivar Improvements

Fantao Zhang[1] and Jiankun Xie[1]

College of Life Sciences, Jiangxi Normal University, China

INTRODUCTION

Rice (Oryza sativa L.) is one of the most important food crops for mankind because it feeds more than half the world population, especially in developing countries (Maclean et al. 2002). Although rice production in the world has increased markedly in the past decades, it is still insufficient to cope with the ever-increasing global demands (Sasaki

Figure 2: Dongxiang wild rice can survive under freezing conditions.

DISEASE RESISTANCE GENES AND QTLS IN WILD RICE

Rice diseases such as blast, bacterial blight and sheath blight are major obstacles for achieving optimal yields. To complement conventional breeding method, molecular or transgenic method represents an increasingly important approach for genetic improvement of disease resistance and reduction of pesticide usage. During the past two decades, a wide variety of genes and mechanisms involved in rice defense response have been identified and elucidated. However, most of the cloned genes confer high level of race specific resistance in a gene-for-gene manner, and the resistance is effective against one or a few related races or strains of the pathogens. The resistance is effective for only few years because the pathogen race or strain keeps changing for survival in nature. Therefore, there is an urgent need to broaden the rice gene pool from diverse resources, of which the wild rice is an ideal option.

Rice Blast Resistance

Rice blast, caused by pathogen Xanthomonas oryzae pv. oryzae, is considered as a major disease of rice because of its wide distribution and destructiveness under favorable conditions. Rice blast causes lesion symptoms on leaves, stems, peduncles, panicles, seeds, and even roots. The potential threat for cropping failure makes this disease ranked among the most devastating diseases in rice. It is reported that the rice blast disease can lead to lose one million hectares annually in China alone (Savary et al. 2000; Khush et al. 2009). Exploitation of resistance gene resources for rice breeding is one of the most important ways to control the disease.

Oryza minuta J. S. Presl ex C. B. Presl is a tetraploid wild rice, highly resistant to rice blast. By genetic analysis, Amante-Bordeos et al (1992) found that the disease resistance was controlled by a single dominant gene, named Pi9. Subsequently, Liu et al (2002) mapped Pi9 in an approximately 100-kb region on chromosome 6, tightly linked with RFLP markers R2132 and RG6. Finally, this broad-spectrum rice blast resistance gene was cloned using a map-based cloning strategy. It turns out that Pi9 encodes a nucleotide-binding site-leucine-rich repeat protein, as a member of a multi-gene family in rice (Qu et al. 2006).

Jeung et al (2007) identified a new gene in the introgression line IR65482-4-136-2-2 that has inherited the resistance gene from an EE genome wild Oryza species, O. australiensis (Acc. 100882). Genetic and molecular analysis localized a major resistance gene, Pi40(t), on the short arm of chromosome 6, within 70-Kb chromosomal region narrowed by two molecular markers RM3330 and S2539. Pi40(t) was validated using the most virulent isolates identified in Korea and the Philippines, suggesting a broad resistance spectrum.

Li et al (2009) evaluated blast resistance for 21 progenies from crossing with common wild rice, and obtained three stably resistance progenies. Preliminary analysis showed that the rice blast resistance was controlled by dominant genes. Geng et al (2008) cloned rice blast resistance gene Pi-ta+ from Jinghong erect type of common wild rice. The Pi-ta+ coding region shares 99.86% and 98.78% of homologous identity with Japanese waxy cultivar Yashiro-mochi and Yuanjiang type of common wild rice, respectively in the corresponding regions. There are four nucleotides difference in the coding region and six in intron of the cloned Pi-ta+ gene, compared with Pi-ta from Yashiro-mochi. The allele in the Jinghong Pi-ta+ gene is very rare in nature, because there is an alanine rather than a serine at position 918 within the predicted amino acid sequence of Pi-ta+. The Pi-ta+ allele can display disease resistance in response to blast pathogens in rice plant cells.

BACTERIAL BLIGHT RESISTANCE

Bacterial blight is caused by Xanthomonas oryzae pv. oryzae. Yield losses due to bacterial blight are variable, heavily dependent on the cultivar used and the environment. In Japan, yield losses ranged typically between 20% and 30% after distribution of high-yielding dwarf varieties (Mew et al. 1993). Among tropical climates, yield losses up to 75% were reported in Indonesia, India, and the Philippines (Nino-Liu et al. 2006). It is of great importance to explore elite bacterial blight resistance genes in rice. By now, a total of 35 related genes have been reported, and nine of which were from wild rice, i.e. Xa21(t), Xa23(t), Xa27(t), Xa29(t), Xa30(t), Xa32(t), xa32(t), Xa35(t) and Xa36(t).

In 1977, Dr. S. Devadath found that a strain of Oryza barthii from Mali is resistant to all the races of bacterial blight in India. Then, Khush et al (1989) found that this strain is akin to O. longistaminata, thus crossed it with IR24, which is susceptible to six races of bacterial blight in Philippines. The F1 was resistant to the six races, thereby showing that the resistance of O. longistaminata was dominant. They designated this gene as Xa21(t). By developing nearly isogenic lines of Xa21(t), Ronald et al (1992) mapped locus Xa21(t) to a region larger than 270 kb on chromosome 11. By positional cloning, Song et al (1995) isolated Xa21(t). The sequence of the predicted protein, which carries both a leucine-rich repeat motif and a serine-threonine kinase-like domain, suggests a role in cell surface recognition of a pathogen ligand and subsequent activation of an intracellular defense response. Furthermore, they demonstrated that the transgenic rice plants carrying the cloned Xa21(t) gene display high levels of resistance to the pathogen.

Xa23(t) was first detected from O. rufipogon by Zhang (2005), showing resistance to race 6 of bacterial blight in the Philippines. Wang et al (2005) constructed a F2 population of JG30/CBB23 for molecule mapping of the Xa23(t) in rice. Based on their previous mapping of Xa23(t) gene, 12 EST markers from Rice Genome Program (RGP) database were surveyed in the susceptible F2 plants and two markers, C189 and CP02662, were found to flank Xa23(t) gene, with genetic distance of 0.8 cM and 1.3 cM, respectively.

Jin et al (2007) identified a rice bacterial blight resistance germplasm (Y238) from the wild rice species Oryza rufipogon, and then they transferred the resistance locus into the cultivated rice to breed near isogenic line. By molecular mapping, the gene Xa30(t) was mapped on the long arm of rice chromosome 11. Linkage analysis revealed that four molecular markers RM1341, V88, C189 and 03STS located on the same side of Xa30(t), with genetic distances of 11.4 cM, 11.4 cM, 4.4 cM and 2.0 cM to the candidate gene, respectively.

Gu et al (2004) performed disease evaluation to a Xa27(t) near-isogenic line, IRBB27 with 35 Xanthomonas oryzae pv. oryzae strains collected from 11 countries. The Xa27(t) gene conferred a high level of resistance to 27 strains and moderate resistance to three strains. Resistance of the Xa27(t) gene was developmentally regulated in IRBB27 and showed semi-dominant or a dosage effect in the cv. CO39 genetic background. Molecular mapping located Xa27(t) within a genetic interval of 0.052 cM, flanked by markers M964 and M1197 and co-segregated with markers M631, M1230, and M449.

Guo et al (2010) transferred a new rice bacterial blight resistance gene Xa35(t) from the wild rice species Oryza minuta (Acc. No. 101133) into Oryza sativa L. (IR24). Through genetic analysis and identification of resistance spectrum, Xa35(t) showed a high level of resistance to PXO61, PXO112 and PXO339, but was susceptible to PXO86 and PXO99 after inoculation with the five strains of Xanthomonas oryzae pv. oryzae. With SSR marker analysis, the Xa35(t) locus was mapped to a 1.80 cM region. This locus was co-segregated with marker RM144, and was 0.7 cM from marker RM6293 on one side and 1.1 cM from marker RM7654 on the other side on rice chromosome 11.

Xa29(t), which was detected from the wild rice Oryza officinalis, has a high resistance to bacterial blight. By molecular mapping, the Xa29(t) gene was mapped within a 1.3 cM region flanked by RFLP markers C904 and R596 on chromosome 1 (Tan et al. 2004). Xa32(t), a bacterial blight resistance gene from Oryzae ustraliensis, was resistant to Xanthmonas oryzae pv. oryzae strains P1 (PXO61), P4 (PXO71), P5 (PXO112), P6 (PXO99), P7 (PXO145), P8 (PXO280), P9 (PXO339), and KX085, but susceptible to P2 (PXO86) and P3 (PXO79). Xa32(t) was mapped within a 2.0 cM interval flanked by two SSR markers RM2064 and RM6293 on the long arm of rice chromosome 11 (Zheng et al. 2009). Miao et al (2010) detected that the rice germplasm C4059 harbored a bacterial blight resistance gene, and designated it tentatively as Xa36(t). By analyzing the mapping populations, the gene Xa36(t) was mapped within a length of 4.5 cM flanked by RM224 and RM2136 on the long arm of rice chromosome 11.

Others

Bacterial leaf streak (BLS) is caused by Xanthomonas oryzae pv. Oryzicola in rice. BLS occurs in Asia and West Africa, and yield losses are up to 30 percent. The symptoms of BLS include translucent interveinal streaks extending to orange lesions which may kill the leaf. Yellowish bacterial exudates may be seen. Bacteria may enter through small wounds on the leaf surface, including insect damage. Plants are susceptible at all stages, but infection is most damaging at the tillering stage. BLS is often prevalent in the rainy season. In order to determine if the resistance genes to the BLS disease were from Guangxi wild rice in China, Huang et al (2008) screened 1655 accessions of Guangxi Oryza rufipogon Griss, and identified 57 (1.87%) accessions to be resistant. In another screening, 15 (48.4%) out of 31 accessions of O. officinalis Wall. ex Watt were resistant.

Sheath blight disease, caused by a soilborne necrotrophic fungus Rhizoctonia solani Kühn, is one of the most important diseases in cultivated rice. This disease was first reported in Japan in 1910 and subsequently discovered worldwide (Rush et al. 1992). At present, rice sheath blight widely occurs in most rice-growing areas, including temperate, tropical and subtropical regions in diverse rice production systems (Lee et al. 1983). Sheath blight disease causes approximately 50% yield reduction in test plots of susceptible cultivars (Savary et al. 1996). To identify resistant germplasm to sheath blight disease, Prasad et al (2008) reported seven Oryza spp. accessions as moderately resistant, three were O. nivara accessions (IRGC104705, IRGC100898, and IRGC104443), O. barthii (IRGC100223), O. meridionalis (IRGC105306), O. nivara/O. sativa (IRGC100943), and O. officinalis (IRGC105979). Greater effort should be paid to search sheath blight resistant germplasm from wild rice and to transfer the resistant genes into the cultivated rice in the future.

INSECT RESISTANCE GENES AND QTLS IDENTIFIED IN WILD RICE

Insects are serious constraints to rice production. In Asia alone, yield loss due to insects has been estimated at about 25% (Savary et al. 2000). Insects not only damage the plant by feeding on its tissue, but also are vectors of devastating rice viruses in many cases. All portions of the plant, from panicle to root, are possibly attacked by various insects. And all growth stages of the rice plant, from the seedling to mature stages, are vulnerable. Even after harvest, the grain in store might face the attack from insects (Cramer et al. 1967). Because the resistance sources in cultivated rice are limited, it is important to keep exploring resistant germplasm from wild rice species for cultivar improvements.

Brown planthopper (BPH) is a destructive insect pest to rice in Asian countries where most rice is produced in the world, including China, India, the Philippines, Japan, Korea, Vietnam, etc (Khush 1984). BPH directly damages the plant phloem by using its piercing-sucking mouthparts, resulting in "hopper burn" in the most serious cases. Furthermore, it is also a vector for rice grassy stunt virus and ragged stunt virus, which may cause further yield losses in many Asian countries (Chelliah et al. 1993). Identification and incorporation of new BPH resistance genes from wild rice into modern cultivars are important breeding strategies to control the damage caused by the BPH.

Ishii et al (1994) found an introgression line from wild species Oryza australiensis resistant to three biotypes of BPH, and named the gene Bph10(t). RFLP analysis resulted in a linkage of the gene Bph10(t) with RG457 on chromosome 12 at a distance of 3.68 +/- 1.29 cM. A BPH biotype-4 resistance gene Bph13(t) was identified from Oryza officinalis Wall. Using RILs where parents "IR50" (cultivar which is susceptible to BPH Biotype-4) and "IR54745-2-21-12-17-6" (a line with Oryza officinalis-derived resistance to BPH biotype-4) are included, Bph13(t) was located on chromosome 3, linked with a RAPD marker AJ09b with the distance of 1.3 cM (Renganayaki et al. 2002).

Later, Jena et al (2006) identified a major BPH resistance gene Bph18(t) from an introgression line (IR65482-7-216-1-2) with wild species Oryza australiensis. Genetic analysis concluded that Bph18(t) is a dominant gene located within a 0.843 Mb physical interval flanked by markers R10289S and RM6869 on the long arm of chromosome 12, where three BAC clones are present.,. Subsequently, Jena et al (2010) successful cloned the Bph18(t) gene. Bph14 is a BPH resistance gene at seedling and maturity stages. Du et al (2009) cloned Bph14 gene to encode a coiled-coil, nucleotide-binding, and leucine-rich repeat (CC-NB-LRR) protein. Sequence comparison indicates that Bph14 carries a unique LRR domain that might function in recognizing the BPH insect invasion and activating the defense response. Bph14 is predominantly expressed in vascular bundles, the site of BPH feeding. Expression of Bph14 activates the salicylic acid signaling pathway and induces callose deposition in phloem cells and trypsin inhibitor production after BPH infestation, thus reducing the BPH feeding to yield low growth rate and longevity of BPH insects.

Rahman et al (2009) conducted a genetic analysis of BPH resistance using an F2 population derived from a cross between an introgression line, IR71033-121-15 from Oryza minuta (Accession number 101141) and a susceptible Korean japonica variety, Junambyeo. Two major QTLs were identified for BPH resistance. One was mapped to 193.4 kb region located on the short arm of chromosome 4, and the other was mapped to a 194.0 kb region on the long arm of chromosome 12.

Abiotic Stress Resistance Genes and QTLs Identified in Wild Rice

Abiotic stresses including high salinity, drought and flood, high and low temperatures are largely limiting productivity of rice crops in large areas of the world. According to Hossain (1996), abiotic stresses affect rice cultivation more than the biotic stresses. Improving the resistance to abiotic stresses will increase agricultural productivity and extend cultivatable areas of rice. There is, therefore, a strong demand for rice cultivars resistant to abiotic stresses.

Based on physiological studies on stress responses, recent progress in plant molecular biology has enabled discovery of many genes involved in stress tolerance. These genes include functional genes which protect the cell (e.g., enzymes for generating protective metabolites and proteins), and regulatory genes which regulate stress response (e.g., transcription factors and protein kinases). Wild rice is the ancestor of cultivated rice, having been an important gene pool due to its survival ability in wild conditions and suffering from natural selection. Therefore, it is of great significance to study genetic basis of abiotic stress resistance as well as to explore new related genes in wild rice.

COLD RESISTANCE

Cold stress is a common problem for rice cultivation, and is a significant factor affecting global food production since cold stress can cause poor germination, slow growth, withering, and anthers injury on rice plants (Andaya et al. 2007). Annually, about 15 million hectares of rice in the world suffered from cold damage (Zhang et al. 2005). In south Asia, about 7 million hectares cannot be planted timely because of the low temperature stress (Sthapit et al. 1998). Consequently, development of rice cultivars with cold tolerance is recognized as one of the important breeding objectives.

Various methods have been adapted to improve rice resistance to low temperature stress (Bertin et al. 1997; Takesawa et al. 2002). With increasing emphasis on F1 hybrid rice production in public institutions and private breeding companies, lots of landraces with diversified genetic background continue to decrease, which makes the genetic base of parental materials become more and more narrower. As a result, development of cultivars for strong cold tolerance becomes increasingly difficult using intra-variation. There is thus an urgent need to study the cold-tolerance character and excavate related genes in wild rice to broaden rice gene pool for developing cold tolerance cultivars.

Genetic analysis of cold tolerance at seedling and/or booting stage has resulted in the identification of many QTLs (Lou et al. 2007; Zeng et al. 2009). Zheng et al (2011) constructed chromosome segment substitution line (CSSL) populations using two core accessions of common wild rice (DP15 and DP30) as donor parents and cultivar 9311 as recipient parent. Thus, they identified cold tolerance QTLs effective at the seedling stage. Two donor lines, DP15 and DP30, are different in the number, location and effect of QTLs for cold tolerance. A total of 19 cold tolerance QTLs were detected, and clustered on chromosome 3 and chromosome 8. The survival rates ranged 8 – 74% after cold treatment among the CSSLs. A major QTL qSCT-3-1 was mapped between SSR markers RM15031 and RM3400, near the centromere of chromosome 3 on the long arm with a distance of 1.8 cM.

Dongxiang wild rice can winter over successful in Wuhan, Hubei province, China, where the lowest temperature can be down to -12C in winter (Liu et al. 2003). In order to transfer cold tolerance gene from Dongxiang wild rice, we have developed introgression lines (ILs) through a backcrossing and single-seed descent program using an elite indica restoring cultivar Xieqingzao B (O. sativa L.) as recipient and Dongxiang wild rice as donor parent (Jian et al. 2011). Analyzing the introgression lines found that the IL5243 and IL5335 were the best for cold tolerant ability (Chen et al. 2013). Genetic analysis using SSR markers further confirmed that a part of alien DNA has been transferred from the common wild rice into IL5243 and IL5335. Therefore, IL5243 and IL5335 might be excellent bridging germplasm for breeding programs to improve cultivar tolerance to cold stress.

SOIL SALINITY RESISTANCE

Soil salinity is one of the major agricultural problems affecting crop productivity worldwide (Rozema et al. 2008). Of the cereals, rice is one of the most salt-sensitive crops (Shelden et al. 2013). The effects of salinity on rice have been reported to reduce seed germination (Hakim et al. 2010), decrease growth and survival of seedlings (Lutts et al. 1995), damage the structure of chloroplasts (Yamane et al. 2008), reduce photosynthesis (Moradi et al. 2007) and inhibit seed set and grain yield (Asch et al. 2000). Improving evaluation methodologies to identify genetic sources and excavating responsible genes for improving cultivar salt resistance is of continuing importance in rice. Oryza coaretata is an Asian wild rice species, occurring mostly in the coastal areas of India. This species is highly resistant to salt because of survival ability in the coastal environments. O. coarctata has some special unicellular hairs (trichomes) on the adaxial surface of leaves. The hairs efficiently maintain a low concentration of toxic salts in the plant tissue (Bal et al. 1986).

Low-Phosphorus Resistance

Phosphorus is one of essential nutritive elements for rice growth and development (Abel et al. 2002). The phosphorus content may be too little in the soil to be able to meet the needs of rice growth. It has been estimated that 5.7 billion hectares of land are deficient in phosphorus worldwide. Phosphorus deficiency is considered as one of the greatest limitations in agricultural production (Schachtman et al. 1998; Lynch et al. 2008).

Chen et al (2011) identified the low-phosphorus resistance ability of Dongxiang wild rice at the seedling stage by using the cultivated low-phosphorus sensitive varieties as the control. The results showed that Dongxiang wild rice has strong low-phosphorus resistance ability. And then, they developed BILs by using Dongxiang wild rice as donor parent and the low-phosphorus sensitive variety Xieqingzao B as recurrent parent. By analyzing the morphological indices, they found that the low-phosphorus resistance lines under low-phosphorus stress had higher values of relative leaf age, relative plant height, relative shoot dry mass, and relative soluble content, but low values of relative yellow leaf number and relative malondialdehyde content, suggesting that the low-phosphorus resistance capability of the low-phosphorus resistance lines was mainly attributed to the high phosphorus utilization efficiency of the lines, namely, low-phosphorus resistance lines had stronger capability in synthesizing dry mass with per unit phosphorus uptake (Chen et al. 2011).

Drought Resistance

Because of global climate warming and increasing scarcity of water resource, drought stress and water scarcity have severely impacted the security of rice production (Farooq et al. 2009). At least 23 million hectares of rice area in Asia are estimated to be drought-prone (Pandey et al. 2005). To date, however, the major challenge for research communities is the relatively limited progress achieved in developing high yielding rice cultivars with drought resistance (Rabello et al. 2008). Therefore, the improvement of drought resistance in newly developed cultivars, for the wide adaptability across rice-growing ecologies, has become a major priority in rice breeding programs. Accordingly,

V

Y

Z

identifying genes from new germplasm resources such as wild rice has become extremely important for drought resistance, which will lay the foundation for utilization of drought resistance gene and genetic improvement of drought resistance (Xie et al. 2004).

Our group has already carried out preliminary experiments for many years on characterization of Dongxiang wild rice for genetic differentiation and conservation, and utilization (Xie et al. 2010). We proved that Dongxiang wild rice has strong drought resistance (Figure 3). Subsequently, Hu et al (2013) constructed BIL population using Indica restorer line R974 (Oryza sativa L.) and Dongxiang wild rice. Using a mixed inheritance model for both major genes and minor genes, they found that the inheritance of drought-resistance at seedling stage was controlled by two independent genes plus polygenes. Therefore, Dongxiang wild rice could be precious resource for genetic improvement of drought resistance in cultivar development.

Figure 3: Dongxiang wild rice has strong drought resistance.

Yield-enhancing QTLs from wild rice

In general, wild rice has smaller seeds and other undesirable traits compared to cultivars, and thus appears not to be appropriate for a donor to enhance yield in cultivars. However, molecular studies have demonstrated that phenotypically poor wild rice contains some genes important for improving cultivar yield (Tanksley et al. 1996). Some wild-QTL alleles are favorable for some traits, but may be associated with deleterious effects on other traits. The positive QTLs from O. rufipogon may be potentially useful for breeding high yield cultivar if the disadvantage linkage drag could be broken through careful selection. In addition, other potentially beneficial QTLs for yield-related traits are often linked to the QTLs conferring negative traits. For example, gpp1.1 with yield increasing effect is closely linked with a negative QTL to increase plant height because this QTL is closely linked to sd1 locus (Cho et al. 2003). Brondani et al (2002) detected specific marker regions to strongly associate with multiple yield-related traits including panicle number, spikelets per panicle, seed set percentage, 100-grain weight, grain yield per plant, filled grain number per panicle and grain yield per panicle.

By using a BC2F5 population derived from the cross between Zhenshan 97 and a wild rice, Wu et al (2012) identified a QTL region flanked by SSR marker RM481 and RM2 on chromosome 7. This QTL has pleiotropic effects on heading date, spikelets per panicle, and grain yield per plant. The alleles from wild rice have increasing effects on these phenotypic traits contributable to grain yield.

Fu et al (2010) developed an advanced backcross population by using an acccssion of common wild rice collected from Yuanjiang County, Yunnan Province, China, as the donor and an elite cultivar 9311 as the recurrent parent. From this population, several QTLs originating from O. rufipogon display beneficial effects for yield-related traits in the 9311 genetic background. In addition, five QTLs controlling yield and its components are newly identified, and they are potentially novel alleles in Yuanjiang common wild rice. Three regions underling significant QTLs for several yield-related traits are detected on chromosome 1 (RM212-RM5362), 7 (RM125-RM1135) and 12 (RM7003-RM277).

Xiao et al (1998) identified two yield-enhancing QTLs, yld1.1 and yld2.1, from O. rufipogon using BC2 populations. QTLs yld1.1 and yld2.1 have been transferred to the elite restorer lines Ce64-7, 9311 and Minghui63 by marker-assisted selection (MAS), and they are confirmed to produce significant yield-enhancing effects in field tests.

Xie et al (2006) fine mapped a yield-enhancing QTL cluster using a BC3F4 population derived from a cross between the Korean japonica cultivar Hwaseongbyeo and O. rufipogon. The cluster contained seven QTLs for 1000-grain weight, spikelets per panicle, grains per panicle, panicle length, spikelet density, heading date and plant height. The alleles from the low-yielding O. rufipogon parent are beneficial in the Hwaseongbyeo background.

Present Problems and Future Directions

As the wild relatives and ancestor of cultivated rice, wild rice carries various characteristics resistant to biotic and abiotic stresses, beneficial agronomic traits, and abundant genetic diversity, which have been lost in the cultivated rice due to breeding activities (Sakai et al. 2010). Thus, it is an extremely important resource for improving important traits in cultivated rice (Xie et al. 2004). However, loss of wild rice genetic diversity was sped up by increasing deterioration of original habitat. For example, the Dongxiang wild rice was sharply reduced from nine populations in nine isolated areas in 1978 to three in 1995 (Hu et al. 2011). The dramatic reduction makes the unique gene pool endangered. Therefore, it is necessary to accelerate a rational conservation for effective utilization of these survived genetic resources.

Breeders have long recognized the intrinsic value of wild rice for improving the traits of modern cultivars. The most successful examples to utilize wild rice in the history of rice breeding include the use of Oryza nivara genes to provide long-lasting resistance to grassy stunt virus (Plucknett et al. 1987), and the use of O. spontanea as a source of wild abortive cytoplasmic male sterility, which has made a cornerstone for today's hybrid rice (Li et al. 1988). However, despite these successes, it is still difficult to utilize wild rice for the improvement of quantitatively inherited traits. Great progress of molecular markers and maps makes it possible to identify individual QTL associated with elite traits from wild rice, which will help transfer the valuable QTLs into modern cultivars to improve their qualities (Tanksley et al. 1996).

Nowadays, QTL studies for mining favorable genes from wild rice species are receiving more and more attentions in global rice community. Several studies have successfully identified and introduced the QTL enhancing alleles from wild rice for yield-related traits into high-yielding elite cultivars (He et al. 2006; Deng et al. 2007; Tan et al.

2008). In addition, some QTLs related to rice quality traits were also detected using wild rice introgression lines (Hao et al. 2006; Garcia-Oliveira et al. 2009). Molecular mapping of these good genes will help discover and make full use of the elite resources of wild species to broaden the genetic base of modern cultivars. However, only a few genes have been cloned from wild rice, and the mechanism for those excellent traits from wild rice are far from being clarified. Cloning more genes from wild rice should be emphasized in the future, which will help make full use of these elite resources more effectively.

In summary, as a rare germplasm resource, wild rice is of great significance to our agricultural heritage and biodiversity protection. Research reveals that wild rice not only has many elite genes which have lost in cultivated rice, but also maintains a greater genetic diversity than cultivated rice. We should use the wild rice to broaden genetic diversity of cultivated rice, by which new cultivars could withstand biotic and abiotic stresses. This is of great significance to assure both high yield and quality in rice production.

REFERENCES

1. Abel S, Ticconi CA, Delatorre CA (2002) Phosphate sensing in higher plants. Physiol Plant 115:1–8
2. Amante-Bordeos A, Sitch LA, Nelson R, Damacio RD, Oliva NP, Aswidinnoor H, Leung H (1992) Transfer of bacterial blight and blast resistance from the tetraploid wild rice *Oryza minuta* to cultivated rice, *Oryza sativa*. Theor Appl Genet 84:345–354
3. Andaya VC, Tai TH (2007) Fine mapping of the *qCTS4* locus associated with seedling cold tolerance in rice (*Oryza sativa* L.). Mol Breeding 20:349–358
4. Asch F, Dingkuhn M, Dorffling K, Miezan K (2000) Leaf K/Na ratio predicts salinity induced yield loss in irrigated rice. Euphytica 113:109–118
5. Bal AR, Dutt SK (1986) Mechanism of salt tolerance in wild rice (*Oryza coarctata Roxb*). Plant and soil 92:399–404
6. Bertin P, Bouharmont J (1997) Use of somaclonal variation and in vitro selection for chilling tolerance improvement in rice. Euphytica 96:135–142

7. Brar DS, Khush GS (1997) Alien introgression in rice. Plant Mol Biol 35:35–47
8. Brar DS, Khush GS (2003) Utilization of wild species of genus *Oryza* in rice improvement. In: Monograph of genus *Oryza*, Nanda, J.S. and S.D. Sharma (eds). Science Publishers, Inc., UK., pp:283–310
9. Brondani C, Rangel PHN, Brondani RPV, Ferreira ME (2002) QTL mapping and introgression of yield-related traits from *Oryza glumaepatula* to cultivated rice (*Oryza sativa*) using microsatellite markers. Theor Appl Genet 104:1192−1203
10. Chelliah S, Bharathi M (1993) Biotypes of the brown planthopper, *Nilaparvata lugens* (homoptera: Delphacidae) – host inxuenced biology and behavior. Chemical ecology of phytopathogous insects. International Science Publishers, New York:133–148
11. Chen XR, Chen M, He HH, Zhu CL, Peng XS, He XP, Fu JR, Ouyang LJ (2011) Identification of low-phosphorus-tolerance in Dongxiang wild rice (*Oryza rufipogon* Griff.). Acta Agriculturae Universitatis Jiangxiensis 33:405–411 (Chinese with English abstract)
12. Chen XR, Chen M, He HH, Zhu CL, Peng XS, He XP, Fu JR, Ouyang LJ (2011) Low phosphorus tolerance and related physiological mechanism of Xieqingzao B // Xieqingzao B / Dongxiang wild rice BC1F9 populations. Chinese Journal of Applied Ecology 22:1169–1174 (Chinese with English abstract)
13. Chen XR, Yang KS, Fu JR, Zhu CL, Peng XS, He P, He HH (2008) Identification and genetic analysis of fertility restoration ability in Dongxiang wild rice (*Oryza rufipogon*). Rice Sci 15:21–28
14. Chen YL, Luo XD, Zhang FT, Dai LF, Hu BL, Xie JK (2013) Cloning and expression analysis of retrotransposon reverse transcriptase in introgression lines from Dongxiang wild rice. Chinese Bulletin of Botany 48:138–144 (Chinese with English abstract)
15. Cho YC, Suh JP, Choi IS, Hong HC, Baek MK, Kang KH, Kim YG, Ahn SN, Choi HC, Hwang HG, Moon HP (2003) QTLs analysis of yield and its related traits in wild rice relative *Oryza rufipogon*. Treat of Crop Res 4:19–29
16. Cramer DA, Barton RA, Shorland FB, Chochanska Z (1967) A

comparison of the effects of white clover (*Trifolium repens*) and of perennial ryegrass (*Lolium perenne*) on fat composition and flavour of lamb. J Agric Sci 69:367−373

17. Deng HB, Deng QY, Chen LY, Yang YS, Liu GM, Zhuang W, Xiong YD (2007) Yield-increasing effect of yield-enhancing QTLs from *O. rufipogon*transferred to 9311, a male parent of medium super hybrid rice. Hybrid Rice 22:49−52 (Chinese with English abstract)
18. Du B, Zhang WL, Liu BF, Hu J, Wei Z, Shi ZY, He RF, Zhu LL, Chen RZ, Han B, He GC (2009) Identification and characterization of *Bph14*, a gene conferring resistance to brown planthopper in rice. PNAS 106: 22163–22168
19. Farooq M, Wahid A, Kobayashi N, Fujita D, Basra SMA (2009) Plant drought stress: effects, mechanisms and management. Agron Sustain Dev 29:185–212
20. Fischer J, Fazey I, Briese R, Lindenmayer DB (2005) Making the matrix matter: challenges in Australian grazing landscapes. Biodivers Conserv 14: 561–78
21. Fu Q, Zhang PJ, Tan LB, Zhu ZF, Ma D, Fu YC, Zhan XC, Cai HW, Sun CQ (2010) Analysis of QTLs for yield-related traits in Yuanjiang common wild rice (*Oryza rufipogon* Griff.). J Genet Genomics 37:147−157
22. Garcia-Oliveira AL, Tan LB, Fu YC, Sun CQ (2009) Genetic identification of quantitative trait loci for contents of mineral nutrients in rice grain. J Integr Plant Biol 51:84−92
23. Geng XS, Yang MZ, Huang XQ, Cheng ZQ, Fu J, Sun T, Li J (2008) Cloning and analyzing of rice blast resistance gene *Pi-ta+* allele from Jinghong erect type of common wild rice (*Oryza rufipogon* Griff) in Yunnan. Hereditas (Beijing) 30(1):109–114 (Chinese with English abstract)
24. Gu K, Tian D, Yang F, Wu L, Sreekala C, Wang D, Wang GL, Yin Z (2004) High-resolution genetic mapping of *Xa27*(*t*), a new bacterial blight resistance gene in rice, *Oryza sativa* L. Theor Appl Genet 108:800–807
25. Guo SB, Zhang DP, Lin XH (2010) Identification and mapping of a novel bacterial blight resistance gene *Xa35*(*t*) originated

from *oryza minuta*. Sci Agri Sin 43:2611–2618 (Chinese with English abstract)

26. Hakim MA, Juraimi AS, Begum M, Hanafi MM, Ismail MR, Selamat A (2010) Effect of salt stress on germination and early seedling growth of rice (*Oryza sativa* L.). Afr J Biotechnol 9:1911–1918
27. Hao W, Jin J, Sun SY, Zhu MZ, Lin HX (2006) Construction of chromosome segment substitution lines carrying overlapping chromosome segments of the whole wild rice genome and identification of quantitative trait loci for rice quality. J Plant Physiol Mol Biol 32:354−362 (Chinese with English abstract)
28. He GM, Luo XJ, Tian F, Li KG, Su W, Zhu ZF, Qian XY, Fu YC, Wang XK, Sun CQ, Yang JS (2006) Haplotype variation in structure and expression of a gene cluster associated with a quantitative trait locus for improved yield in rice. Genome Res 16: 618−626
29. Hossain M (1996) Economic prosperity in Asia: Implication for rice research. IRRI. Rice Genetics III. Proceedings of the third international rice genetics symposium. 16–20 Oct., 1995, Manila, Philippines
30. Hu BL, Fu XQ, Zhang T, Wan Y, Li X, Huang YH, Dai LF, Luo XD, Xie JK (2011) Genetic analysis on characteristics to measure drought resistance using Dongxiang wild rice (*Oryza rufupogon* Griff.) and its derived backcross inbred lines population at seedling stage. Agricultural Sciences in China 10:1653–1664
31. Hu BL, Yang P, Wan Y, Li X, Luo SY, Luo XD, Xie JK (2013) Comprehensive assessment of drought resistance of BILs population derived from Dongxiang wild rice (*Oryza rufupogon* Griff.) at seedling stage and its genetic analysis. Journal of Plnat Genetic Resources 14:249–256 (Chinese with English abstract)
32. Huang DH, Cen ZL, Liu C, He WN, Chen YZ, Ma ZF, Yang L, Wei SL, Liu YL, Huang SL, Yang XQ, Li RB (2008) Identification and genetic analysis of resistance to bacterial leaf streak in wild rice. Journal of plant genetic resources 9:11−14 (Chinese with English abstract)
33. Ishii T, Brar DS, Multani DS, Khush GS (1994) Molecualar tagging of genes for brown lanthopper resistance and earliness introgressed from *Oryza australiennsis* into cultivated rice, *O. sativa*. Genome 37:217−221

34. Jena KK, Jeung JU, Lee JH, Choi HC, Brar DS (2006) High-resolution mapping of a new brown planthopper (BPH) resistance gene *Bph18(t)*, and marker-assisted selection for BPH resistance in rice (*Oryza sativa* L.). Theor Appl Genet 112:288–297

35. Jena KK, Kim SM (2010) Current status of brown planthopper (BPH) resistance and genetics. Rice 3:161–171

36. Jeung JU, Kim BR, Cho YC, Han SS, Moon HP, Lee YT, Jena KK (2007) A novel gene, *Pi40(t)*, linked to the DNA markers derived from NBS-LRR motifs confers broad spectrum of blast resistance in rice. Theor Appl Genet 115:1163–1677

37. Jian SR, Wan Y, Luo XD, Fang J, Chu CC, Xie JK (2011) Genetic analysis of cold tolerance at the seedling stage in Dongxiang wild rice (*Oryza rufipogon*). Chinese Bulletin of Botany 46: 21–27 (Chinese with English abstract)

38. Jin XW, Wang CL, Yang Q, Jiang QX, Fan YL, Liu GC, Zhao KJ (2007) Breeding of near-isogenic line CBB30 and molecular mapping of *Xa30(t)*, a new resistance gene to bacterial blight in rice. Sci Agri Sin 40:1094–1100 (Chinese with English abstract)

39. Khush GS (1984) Breeding rice for resistance to insects. Protection Eco 7:147–165

40. Khush GS, Mackill DJ, Sidhu GS (1989) Breeding rice for resistance to bacterial blight. In: Bacterial blight of rice. International Rice Research Institute, Manila, Philippines:207–217

41. Khush GS (1997) Origin, dispersal, cultivation and variation of rice. Plant Mol Biol 35: 25–34

42. Khush GS, Jena KK (2009) Current status and future prospects for research on blast resistance in rice (*Oryza sativa* L.). Advances in genetics, genomics and control of rice blast disease. Berlin: Springer:1–10

43. Lee FN, Rush MC (1983) Rice sheath blight: a major rice disease. Plant Dis 67:829–832

44. Li SH, Yang QW, Zheng GB, Li H, Lu XL, Han GM (2009) The genetic analysis on resistance to rice blast of rice progeny introduced with wild rice genes. Seed 28:36–40 (Chinese with English abstract)

45. Li Z, Zhu Y (1988) Rice male sterile cytoplasm and fertility restoration. In: Hybrid Rice. International Rice Research Institute. Manila, Philippines:85–102

46. Liu FX, Sun CQ, Tan LB, Fu YC, Li DJ, Wang XK (2003) Identification of QTL for cold tolerance at booting and flowering stage in Jiangxi Dongxiang wild rice. Chin Sci Bull 48:1864–1867 (Chinese with English abstract)

47. Liu G, Lu G, Zeng L, Wang GL (2002) Two broad-spectrum blast resistance genes, *Pi9*(*t*) and *Pi2*(*t*), are physically linked on rice chromosome 6. Mol Genet Genomics 267:472–80

48. Lou QJ, Chen L, Sun ZX, Xing YZ, Li J, Xu XY, Mei HW, Luo LJ (2007) A major QTL associated with cold tolerance at seedling stage in rice (*Oryza sativa* L.). Euphytica 158: 87–94

49. Lutts S, Kinet J M, Bouharmont J (1995) Changes in plant response to NaCl during development of rice (*Oryza sativa* L.) varieties differing in salinity resistance. J Exp Bot 46:1843–1852

50. Lynch JP, Brown KM (2008) Root strategies for phosphorus acquisition. Plant Ecophysiol 7:83–116

51. Maclean JL, Dawe DC, Hardy B, Hettel GP (eds) (2002) Rice almanac (Third Edition). Philippines, IRRI, WARDA, CIAT and FAO

52. Mew TM, Alvarez AM, Leach JE, Swings J (1993) Focus on bacterial blight of rice. Plant Dis 77:5−12

53. Miao LL, Wang CL, Zheng CK, Che JY, Gao Y, Wen YC, Li GQ, Zhao KJ (2010) Molecular mapping of a new gene for resistance to rice bacterial blight. Sci Agri Sin 43: 3051−3058 (Chinese with English abstract)

54. Moradi F, Ismail AM (2007) Responses of photosynthesis, chlorophyll fluorescence and ROS-scavenging systems to salt stress during seedling and reproductive stages in rice. Annals of Botany 99:1161–1173

55. Nino-Liu DO, Ronald PC, Bogdanove AJ (2006) *Xanthomonas oryzae* pathovars: model pathogens of a model crop. Mol Plant Pathol 7:303−324

56. Pandey S, Bhandari H, Sharan R, Naik D, Taunk SK, Sastri ASRAS (2005) Economic costs of drought and rainfed rice farmers'

coping mechanisms in eastern India, Final report. International Rice Research Institute, Los Baños, Philippine.

57. Plucknett DL, Smith NJH, Williams JT, Anishetty NM (1987) Gene banks and the world's food. Princeton University Press, Princeton, New Jersey

58. Prasad B, Eizenga GC (2008) Rice sheath blight disease resistance identified in *Oryzas pp. accessions*. Plant Dis 92:1503–1509

59. Qu SH, Liu GF, Zhou B, Bellizzi M, Zeng LR, Dai LY, Han B, Wang GL (2006) The broad-spectrum blast resistance gene *Pi9* encodes a nucleotide-binding site-leucine-rich repeat protein and is a member of a multigene family in rice. Genetics 172:1901–1914

60. Rabello AR, Guimaraes CM, Rangel PH, da Silva FR, Seixas D, de Souza E, Brasileiro AC, Spehar CR, Ferreira ME, Mehta A (2008) Identification of drought-responsive genes in roots of upland rice (*Oryza sativa* L). BMC Genomics 9:485

61. Rahman ML, Jiang W, Chu SH, Qiao Y, Ham TH, Woo MK, Lee J, Khanam MS, Chin JH, Jeung JU, Brar DS, Jena KK, Koh HJ (2009) High-resolution mapping of two brown planthopper resistance genes, *Bph20*(*t*) and *Bph21*(*t*), originating from *Oryza minuta*. Theor Appl Genet 119:1237–1246

62. Renganayaki K, Fritz AK, Sadasivam S, Pammi S, Harrington SE, McCouch SR, Kumar SM, Reddy AS (2002) Mapping and progress toward map-based cloning of brown planthopper biotype-4 resistance gene introgressed from *Oryza officinalis* into cultivated rice, *O. sativa*. Crop Science 42:2112–2117

63. Ronald PC, Albano B, Tabien R, Abenes L, Wu KS, McCouch S, Tanksley SD (1992) Genetic and physical analysis of the rice bacterial blight disease resistance locus, *Xa21*. Mol Gen Genet 236:113–120

64. Rozema J, Flowers T (2008) Crops for a salinized world. Science 322:1478–1480

65. Rush MC, Lee FN (1992) Sheath blight. Pages 22–23 in: Compendium of rice diseases. R. K. Webster and P. S. Gunnell, eds. American Phytopathological Society, St. Paul, MN

66. Sakai H, Itoh T (2010) Massive gene losses in Asian cultivated rice unveiled by comparative genome analysis. BMC Genomics 11:121

67. Sasaki T, Burr B (2000) International Rice Genome Sequencing Project: The effort to completely sequence the rice genome. Curr Opin Plant Biol3:138–141
68. Savary S, Mew TW (1996) Analyzing crop losses due to *Rhizoctonia solani*: Rice sheath blight, a case study. In: *Rhizoctonia* Species: Taxonomy, Molecular Biology, Ecology, Pathology and Disease Control. B. Sneh, S. Jabajihare, S. Neate, and G. Dijst, eds. Kluwer Academic Publisher, Dordrecht, The Netherlands:237–245
69. Savary S, Willocquet L, Elazegui FA, Castilla N, Teng PS (2000) Rice pest constraints in tropical Asia: quantification of yield losses due to rice pests in a range of production situations. Plant Dis 84:357–369
70. Schachtman DP, Reid RJ, Ayling SM (1998) Phosphorus uptake by plants: from soil to cell. Plant Physiol 116:447–453
71. Shelden MC, Roessner U (2013) Advances in functional genomics for investigating salinity stress tolerance mechanisms in cereals. Frontiers in plant science 4:123
72. Song WY, Wang GL, Chen LL, Kim HS, Pi LY, Holsten T, Gardner J, Wang B, Zhai WX, Zhu LH, Fauquet C, Ronald P (1995) A receptor kinase-like protein encoded by the rice disease resistance gene, *Xa21*. Science 270:1804–1806
73. Song ZP, Li B, Chen JK, Lu BR (2005) Genetic diversity and conservation of common wild rice (*Oryza rufipogon*) in China. Plant Spec Biol 20:83–92
74. Stebbins GL (1981) Why are there so many species of flowering plants? Bioscience 31:573–577
75. Sthapit BR, Witcombe JR (1998) Inheritance of tolerance to chilling stressing rice during germination and plumule greening. Crop Sci 38:660–665
76. Sun CQ, Wang XK, Li ZC, Yoshimura A, Iwata N (2001) Comparison on the genetic diversity of common wild rice (*Oryza rufipogon* Griff.) and cultivated rice (*O. sativa* L.) using RFLP markers. Theor Appl Genet 102:157–162
77. Takesawa T, Ito M, Kanzaki H, Kameya N, Nakamura I (2002) Over-expression of ζ glutathione S-transferase in transgenic rice enhances germination and growth at low temperature. Mol Breed 9:93–101

78. Tan GX, Ren X, Weng QM, Shi ZY, Zhu LL, He GC (2004) Mapping of a new resistance gene to bacterial blight in rice line introgressed from *Oryza officinalis*. Acta Genetica Sinica 31:724–729 (Chinese with English abstract)

79. Tan LB, Zhang PJ, Liu FX, Wang GJ, Ye S, Zhu ZF, Fu YC, Cai HW, Sun CQ (2008) Quantitative trait loci underlying domestication and yield-related traits in *Oryza rufipogon* × *Oryza sativa* advanced backcross population. Genome 51:692−704

80. Tanksley SD, Grandillo S, Fulton TM, Zamir D, Eshed Y, Petiard V, Lopez J, Beck-Bunn T (1996) Advanced backcross QTL analysis in a cross between an elite processing line of tomato and its wild relative L. pimpinellifolium. Theor Appl Genet 92:213–224

81. Tanksley SD, Nelson JC (1996). Advanced backcross QTL analysis: a method for the simultaneous discovery and transfer of valuable QTLs from unadapted germplasm into elite breeding lines. Theor Appl Genet 92:191–203

82. Vaughan DA (1994) The wild relatives of rice: a genetic resources guide book. International Rice Research Institute, Los Banos, The Philippines

83. Wang CL, Qi HX, Pan HJ, Li JB, Fan YL, Zhang Q, Zhao KJ (2005) EST-markers flanking the rice bacterial blight resistance gene *Xa23* and their application in marker-assisted selection. Sci Agri Sin 38:1996–2001 (Chinese with English abstract)

84. Wu B, Han ZM, Li ZX, Xing YZ (2012) Discovery of QTLs increasing yield related traits in common wild rice. Yi Chuan 34:215−22 (Chinese with English abstract)

85. Xiao JH, Li JM, Grandillo S, Ahn SN, Yuan LP, Tanksley SD, McCouch SR (1998) Identification of trait-improving quantitative trait loci alleles from a wild rice relative, *Oryza rufipogon*. Genetics 150:899–909

86. Xie JK, Agrama HA, Kong D, Zhuang J, Hu B, Wan Y, Yan W (2010) Genetic diversity associated with conservation of endangered Dongxiang wild rice (*Oryza rufipogon*). Genet Resour Crop Evol 57:597–609

87. Xie JK, Hu BL, Wan Y, Zhang T, Li X, Liu RL, Huang YH, Dai LF, Luo XD (2010) Comparison of the drought resistance characters at Seedling Stage between Dongxiang Common Wild rice (*Oryza*

rufipogon Griff.) and cultivars (*Oryza sativa* L.). Acta Ecologica Sinica 30:1665–1674 (Chinese with English abstract)

88. Xie JK, Kong XL, Bao JS, Wan Y (2004) Recent advances in molecular mapping and cloning of useful genes from wild rice and their application in breeding. Hereditas (Beijing) 26:115–121 (Chinese with English abstract)
89. Xie X, Song MH, Jin F, Ahn SN, Suh JP, Hwang HG, McCouch SR (2006) Fine mapping of a grain weight quantitative trait locus on rice chromosome 8 using near-isogenic lines derived from a cross between *Oryza sativa* and *Oryza rufipogon*. Theor Appl Genet 113:885–894
90. Yamane K, Kawasaki M, Taniguchi M, Miyake H (2008) Correlation between chloroplast ultrastructure and chlorophyll fluorescence characteristics in the leaves of rice (*Oryza sativa* L.) grown under salinity. Plant Prod Sci 11:139–145
91. Zeng YW, Yang SM, Cui H, Yang XJ, Xu LM, Du J, Pu XY, Li ZC, Cheng ZQ, Huang XQ (2009) QTLs of cold tolerance-related traits at the booting stage for NIL-RILs in rice revealed by SSR. Genes Genom 31:143–154
92. Zhang Q (2005) Highlights in identification and application of resistance genes to bacterial blight. Chinese J Rice Sci 19:453–459 (Chinese with English abstract)
93. Zhang ZH, Li S, Wei L, Wei C, Zhu YG (2005) A major QTL conferring cold tolerance at the early seedling stage using recombinant inbred lines of rice (*Oryza sativa* L.). Plant Sci 168:527–534
94. Zheng CK, Wang CL, Yu YJ, Liang YT, Zhao KJ (2009) Identification and molecular mapping of *Xa32(t)*, a novel resistance gene for bacterial blight (*Xanthomonas oryzae* pv. *oryzae*) in rice. Acta Agronomica Sin 35:1173–1180 (Chinese with English abstract)
95. Zheng JX, Ma ZF, Song JD, Liu C, Li YT, Huang DH, Wei SL, Zhang YX, Mi K, Huang JY, Chen M, Meng JR, Li RB, Chen BS (2011) Identification and mapping of QTLs for cold tolerance at the seedling stage in common wild rice (*Oryza rufipogon*). Chinese J Rice Sci 25:52–58 (Chinese with English abstract)

Citations

CHAPTER 1

Yanchang Luo, Sabaruddin Zakaria, Bakhtiar Basyah, Tingchen Ma, Zefu Li, Jianbo Yang and Zhongchao Yin, Marker-Assisted Breeding of Indonesia Local Rice Variety Siputeh for Semi-Dwarf Phonetype, Good Grain Quality and Disease Resistance to Bacterial Bligh, doi:10.1186/s12284-014-0033-2.

CHAPTER 2

Nourollah Ahmadi, Alain Audebert, Malcolm J Bennett, Anthony Bishopp, Antonio Costa de Oliveira, Brigitte Courtois, Abdala Diedhiou, Anne Diévart, Pascal Gantet, Alain Ghesquière, Emmanuel Guiderdoni, Amelia Henry, Yoshiaki Inukai, Leon Kochian, Laurent Laplaze, Mikael Lucas, Doan Trung Luu,Baboucarr Manneh, Xiaorong Mo, Raveendran

Muthurajan, Christophe Périn,Adam Price, Sabariappan Robin, Hervé Sentenac, Bassirou Sine, Yusaku Uga, Anne Aliénor Véry, Matthias Wissuwa, Ping Wu and Jian Xu, The Roots of Future Rice Harvests, doi:10.1186/s12284-014-0029-y

CHAPTER 3

Aakash Chawade, Angelica Lindlöf, Björn Olsson and Olof Olsson, Global Expression Profiling of Low Temperature Induced Genes in the Chilling Tolerant Japonica Rice Jumli Marshi, DOI: 10.1371/journal.pone.0081729

CHAPTER 4

Chung D Mai, Nhung TP Phung, Huong TM To, Mathieu Gonin, Giang T Hoang, Khanh L Nguyen, Vinh N Do, Brigitte Courtois and Pascal Gantet, Genes controlling root development in rice, doi:10.1186/s12284-014-0030-5

CHAPTER 5

Yang-Seok Lee, Dong-Yeon Lee, Lae-Hyeon Cho and Gynheung An, Rice miR172 induces flowering by suppressing OsIDS1 and SNB, two AP2 genes that negatively regulate expression ofEhd1 and florigens, doi:10.1186/s12284-014-0031-4

CHAPTER 6

Chao Yu, Huamin Chen, Fang Tian, Jan E Leach and Chenyang He*, Differentially-Expressed Genes in Rice Infected by Xanthomonas Oryzae pv. Oryzae Relative to a Flagellin-Deficient Mutant Reveal Potential Functions of Flagellin in Host–Pathogen Interactions, doi:10.1186/s12284-014-0020-7

CHAPTER 7

Backki Kim, Dong-Gwan Kim, Gileung Lee, Jeonghwan Seo, Ik-Young Choi, Beom-Soon Choi, Tae-Jin Yang, Kwang Soo Kim, Joohyun Lee, Joong Hyoun Chin and Hee-Jong Koh, Defining the Genome Structure of Tongil rice, an important Cultivar in the Korean Green Revolution, doi:10.1186/s12284-014-0022-5

CHAPTER 8

Ehsan Shakiba and Georgia C. Eizenga (2014), Unraveling the Secrets of Rice Wild Species, Rice - Germplasm, Genetics and Improvement, Dr. Wengui Yan (Ed.), ISBN: 978-953-51-1240-2, InTech, DOI: 10.5772/58393.

CHAPTER 9

Fantao Zhang and Jiankun Xie (2014). Genes and QTLs Resistant to Biotic and Abiotic Stresses from Wild Rice and Their Applications in Cultivar Improvements, Rice - Germplasm, Genetics and Improvement, Dr. Wengui Yan (Ed.), ISBN: 978-953-51-1240-2, InTech, DOI: 10.5772/56825.

Index

V

Y

Z